Issues In Advanced Television Technology

ISSUES IN ADVANCED TELEVISION TECHNOLOGY

S. Merrill Weiss

Focal Press
Boston   Oxford   Johannesburg   Melbourne   New Delhi   Singapore

Focal Press is an imprint of Butterworth–Heinemann

 A member of the Reed Elsevier group

Copyright © 1996 by S. Merrill Weiss

Recognizing the importance of preserving what has been written, Butterworth–Heinemann prints its books on acid-free paper whenever possible.

Library of Congress Cataloging-in-Publication Data

Weiss, S. Merrill.
    Issues in advanced television technology / S. Merrill Weiss.
       p.   cm.
    Drawn from the pages of the author's popular "Advanced television"
column in TV technology magazine.
    Includes index.
    ISBN 0-240-80250-0 (pbk. : alk. paper)
    1. High definition television—Popular works.   I. Title.
TK6679.W45   1996
384.55'2—dc20   96-1161
                                                        CIP

British Library Cataloguing-in-Publication Data
A catalogue record for this book is available from the British Library.

The publisher offers special discounts on bulk orders of this book.
For information, please contact:
Manager of Special Sales
Butterworth–Heinemann
313 Washington Street
Newton, MA 02158–1626
Tel: 617-928-2500
Fax: 617-928-2620

For information on all Focal Press publications available, contact our World Wide Web home page at: http://www.bh.com/bh/

10 9 8 7 6 5 4 3 2 1

Printed in the United States of America

To my wife, Carol,
one of the most understanding people I know,
without whose patience and support
the original series would not have been written
and this book could not have been produced.

Contents

List of Figures

List of Tables

Introduction

Advanced television technology (small a, small t's) has been a passion of mine for nearly twenty years. Ever since I was first introduced into the world of the SMPTE standards activities that were trying to figure out how to harness digital techniques and apply them to television at the systems level, I have been hooked. During the time since then, I have been privileged to participate in and to contribute in some small way to the development of those technologies.

Advanced Television Technology (capital a, capital t's) has been my area of particular endeavor for nearly ten years. It is largely how I make my living, and it is an area of activity that I have seen come to have increasing influence on the course of the television industry, taken in its broadest sense, and on the viewing public, too, as time has gone by. I now spend most of my work effort in helping clients understand these new technologies, strategize how to use them in addressing their business needs, select among the many offerings and offerers, and design their systems to maximize their returns on their technology investments.

Ever since that first introduction to the SMPTE standards process, I have seen participation in it as a way to give something back to the industry that has provided me my livelihood all this time. I continue active involvement up to the present for this reason (not to mention the personal rewards that come from working with others to develop methods and techniques that have become pervasive throughout both the television business and many other related industries).

For similar reasons, after having been coaxed for a number of years by the publisher and editors of *TV Technology* magazine to share my experience by writing regularly for them, I began a monthly column on Advanced Television with the July issue of 1992. This book represents a collection of those articles up through the one published in the issue of December, 1995.

In my writings, both in the column and in this book, my intent is to share what I have learned with those who may have need of it even more than I do but who do not have the opportunity for participation that I have had. My intended audience, for this book as for the articles originally, is those people involved in the application of television technology who will someday have to apply these new methods in their everyday work situations and the managers of television and related fields who will need to make decisions about the use of these technologies and who will therefore need to understand them to some extent.

Often, meeting these objectives means taking fairly complex and decidedly obscure material and distilling it to its essence, then finding a way to express it that makes it accessible to those who have not studied the technology in great detail. Commonly, the information needed is buried in some very arcane standards documents. Frequently, it takes a significant amount of study to sort out just what is needed to explain the subject without making the explanation overly complicated. With any luck, the end result of these efforts has fulfilled the objectives reasonably well.

The general approach and tone that I try to adopt in the articles is one of the reader and myself jointly exploring some facet or another of the new world of Advanced Television technology that is opening up before us. I hope readers (you) get the sense that I am conversing with them (you) one-on-one, not in any way lecturing. I am trying to share with you experiences that I have had and developments that I have observed that you yourself might have liked to have had and seen had you only been given the opportunity. Thus I tend to use the word "we" a great deal to convey this outlook.

This book has come about because many readers of the magazine column requested it. Some have shown me their files with all the articles carefully clipped out, preserved, and grouped by subject. Others have told me how they discovered the series well after the beginning and were unable to obtain back copies of the magazine because other latecomers, also looking to catch up, had cleaned out the stock of back issues. Still others have asked me if there was some way that I might publish an index to the articles. The volume in your hands is the answer to all these needs and requests.

You will note throughout this book that the terrestrial broadcast version of Advanced Television gets more than its share of attention. This results from the fact that the development of the terrestrial broadcasting standard has been largely an open process. Developments in other areas tend to happen behind closed doors between specific companies, not as an industry-wide effort. Information about such efforts tends to be circumscribed by non-disclosure agreements, by many of which I am personally obligated in my work. Thus it is much easier to write about the broadcasting case, which makes the trade news anyway, and to consider it an example of how the technology works in most of the other applications as well.

In addition to adding the usual formalities that appear in a book like this of a table of contents, a list of figures, a list of tables, and an index, I have tried to do some updating to some of the articles as I was putting the book together. Thus there are now included in the chapter on MPEG-2 profiles and levels some details on the 4:2:2 profile that did not exist at the time it was originally written. Various places throughout the book, I have added comments in the text or footnotes that provide status updates on work that was still in process or decisions that were still to be reached at the time the articles were originally written. The footnotes are all additions to the original material. New comments in the text are generally indicated with the use of [square brackets]. They are often used to insert years to indicate when things happened that were not originally identified by year because the article was written contemporaneously with the events. When there was already a parenthetical comment in the original that needed to be updated, I just changed it.

There is a great temptation in compiling a book such as this to rewrite large sections in order to make a truly coherent work out of the material. I have resisted that temptation. Doing so would, I think, have removed a lot of the flavor of spontaneity and the conversational tone from the resulting work. This way, the chapters are shorter — probably just long enough to put you to sleep if you read this in bed. In keeping the shorter structure, I have also retained much of the transitional structure that helps tie articles in mini-series together from month-to-month. Thus you will still find the previews of coming attractions that appear at the end of articles in mini-series and the quick recaps that tend to begin the next articles in such series. The recaps at the start of chapters in this book will help remind you where you left off when you fell asleep the night before.

Thank you for joining me on this journey into the world of television to come. I hope you enjoy our adventure together as much as I already have.

ACKNOWLEDGEMENTS

There are many people who have contributed to the original series of articles, and there are more who have contributed to the publishing of this book. I would like to acknowledge them all; if any are left out, it is solely due to my shortfall in memory, and I apologize in advance.

The original instigators of this effort are Steve Dana, the persistent publisher of *TV Technology*, and the associate publisher, Carmel King. Editors of the magazine over the years have been Marlene Lane, Richard Farrell, and Mary Ann Dorsie. They have all been as generous as they were able in extending deadlines to the limit, to which I always pushed them because of the way I always try to do too much – in business and in life in general.

Many industry luminaries have given graciously of their time to provide information or to read the articles to make sure that I got things right. (This is where I will probably leave out someone deserving of recognition.) Those who come to mind as having helped at one time or another include: Ken Ainsworth,

Stan Baron, Lynn Claudy, Jules Cohen, Aldo Cugnini, Carl Eilers, David Fibush, Jack Fuhrer, John Henderson, Bob Hopkins, Bernie Lechner, Woo Paik, Larry Pearlstein, Bruce Penney, Glenn Reitmeier, Charles Rhodes, Bob Ross, Craig Tanner, Victor Tawil, Craig Todd, and Tony Uyttendaele.

Editors at Focal Press who have been most helpful, once again by pushing production deadlines to the limit, and by being encouraging and supportive have been Marie Lee and Maura Kelly. They have both offered suggestions that have contributed to the quality of the result you see before you.

Finally, I would like to acknowledge the many readers of the *TV Technology* column who have encouraged me to stay the course when there seemed not to be the time to carry on. Comments from industry figures whom I respect greatly that they read the column regularly, find it helpful, and would not like to see it end have provided the energy to keep going. Words of appreciation from readers whom I have never before met, combined with requests for the material in book form have led to this result. I trust that you will find it worthy.

PART 1

Where to Begin?...
Where Have We Been?...
Where Are We Going?...

The realm of Advanced Television is so vast that it is often difficult to tell where it begins and where it ends. Even narrowing the area of interest to just the technology of Advanced Television does not help too much because the issues surrounding it include far more than mere technological innovation. They include the efforts of the people who do the invention. They include the interests of the organizations that support that work or that want to employ the fruits of the inventors' labors. They include the financial interests of the players in the field. They include the interests of the public at large, as sometimes expressed, although not always well, by various individuals and agencies in government. And they include a significant amount of techno-politics, in addition to the ordinary economic and governance politics that impinge on any major economic activity.

To get one's bearings in such an environment it helps to know where you began, where you have been, and where you are going. To get us going, Part 1 collects the first article in the series, one in the middle, and the most recent one that is included in this book. Each answers one of the questions posed by the title of the section.

Thus we start by answering the question "Where to begin?" with the introductory article that set the stage for all the rest to come, "Making Sense of It All." It attempts to define what it is that we will be discussing in the remainder of the book (and in the series of articles). It defines some of the major terms. And it explains the fundamental model of compression systems that will underlie all of our other discourse on the subject.

"Where have we been?" is answered by a chapter on one of the more significant events in the development of broadcast advanced television, the FCC Advisory Committee meeting that confirmed the digital nature of the medium. Chapter 2 also recognizes some of the early contributors to the effort and predicts the formation of the Grand Alliance.

Finally, we look ahead by discussing "Where are we going?" The third chapter takes note of the completion of the work of the Advisory Committee and the making of a recommendation to the FCC for the broadcast standard. It points out all of the work yet to be done to reach a practical implementation. It should be noted that much of the same work is also needed and under way in the many non-broadcast industries that share the same technology.

Making Sense of It All

"The Russians Are Coming! The Russians Are Coming!" was a 1965 motion picture, starring Alan Arkin and Carl Reiner, about a Gloucester (Nantucket) Island community that goes a bit crazy when a Russian submarine runs aground nearby. This occurs at a time when there is only lack of knowledge of real Russians and a consequent ignorant fear of them. Having been forced into proximity with a feared enemy, the people on both sides in the movie get to know one another and find out that there really isn't so much to fear once there is knowledge.

For some in the television industry today, "HDTV Is Coming! ATV Is Coming!" might be indicative of their feelings. As in the 60's regarding the Russians, there is a lack of widespread knowledge of HDTV and ATV and, in some quarters, a fear of them that results from the ignorance. Some parts of the television industry will welcome ATV/HDTV with open arms, while others continue to fear what is coming. But real knowledge of the subject can assuage any concerns by those who are fearful and make easier the meeting even for those with open arms.

This column is intended to provide the background information that will help make implementing HDTV and ATV not only something not to be feared but, ideally, something to be enjoyed and to which we can look forward. The column will explore Advanced Television in its broadest sense, including in addition to High Definition Television the related technologies of compressed, multi-channel standard television. Thus we will include the future distribution technologies of all of the visual electronic media — terrestrial broadcast, cable, direct broadcast satellite, wireless cable, and packaged programming (tape and disc) — and the production systems required to feed them.

First published July, 1992. Number 1 in the series.

This column will be both technical and non-technical. Sometimes it will explore technical matters and issues, with an eye to helping non-technical readers understand some of the technology. Sometimes it will explore non-technical matters and issues, with an eye to helping technical readers see that there is far more to Advanced Television than just the technologies involved. At all times, this column will try to provide context for the news stories on ATV that now appear in every issue of every trade publication but often without adequate framework.

WHAT IS ADVANCED TELEVISION?

Advanced Television, as we will use the term here, is an agglomeration of techniques, based largely on digital signal processing and transmission, that permits far more program material to be carried through channels than existing analog systems can manage. In this sense, HDTV (High Definition Television) is a subset of ATV (Advanced Television). ATV is also a new way to look at the businesses in which we are involved — one that allows for a wide range of new or improved services impossible with earlier forms.

Advanced Television has another, somewhat more limited, set of meanings derived from the FCC process to establish an Advanced Television Service for terrestrial broadcast television. While the FCC activity is currently trying to define exactly what it means by ATV, for use in setting certain aspects of the Rules, it generally refers to ATV in terms of the characteristics of HDTV and certain related lower levels of performance. We shall be at pains throughout this column to differentiate between ATV in the more targetted FCC sense and in the broader sense intended here.

It is important to note that, while ATV implies the use of many advanced technologies, the term ATV as we use it to define the scope of this column does not automatically signify improved picture or sound performance. Those are things that can be accomplished with ATV in systems designed for such purposes, but ATV technologies can just as readily provide other features — for example, carrying ten somewhat lower-quality signals where only one could exist previously, or permitting ghost cancellation for ordinary NTSC signals. In each case, however, the new features derive from the use of digital techniques of one form or another.

WHAT IS HIGH DEFINITION TELEVISION?

HDTV is generally defined as a system offering as a minimum certain specific features and characteristics. These are:

1.   Wide aspect ratio (now agreed as 16:9, or 1.778:1).
2.   Effectively doubled horizontal resolution (compared to existing systems).
3.   Effectively doubled vertical resolution (compared to existing systems).
4.   Absence of encoding/decoding artifacts (requires component operation).
5.   Compact disc quality stereo sound.

Although not originally part of the formal definition of HDTV, the word "effectively" is applied to items 2 and 3 above to open the discussion to include systems that have somewhat less actual resolution but are claimed to have the same visual impact as systems with truly doubled resolution by virtue of progressive (non-interlaced) scanning and increased frame rates. We will deal with these issues in a future installment of this column.

HDTV and its techniques will be one of the most important areas examined in this column. As we shall see in future issues, the technology applied to make HDTV transmittable in existing 6 MHz channels is essentially the same as the technology necessary for multi-channel operation in those same channels. By studying one, we can learn about the other. We will use this fact to advantage as we progress through examinations of various systems.

WHAT ARE MULTI-CHANNEL SYSTEMS?

Multi-channel systems are the other (besides HDTV) important subset of uses of Advanced Television technology. As mentioned previously, they share the same suite of techniques used to permit HDTV transmission in a single 6 MHz channel. Multi-channel systems are sometimes referred to as "NTSC Compression" systems, but this is a misnomer because they quite often do not carry true NTSC signals. Instead, they normally begin with component signals at the 525-line rate of NTSC, since it is usually easier to compress signals that have not undergone the analog compression of NTSC and that do not have the artifacts of NTSC. While we may use the shorthand of "NTSC Compression" at times in future articles, we will try to explain what is meant the first time the expression is used in each case.

ADVANCED TELEVISION SYSTEMS — A GENERIC OVERVIEW

ATV systems have certain generic structures at an overview level that can help us understand the relationships of the various parts as we look at those parts in greater

• Compression/Transmission/Storage

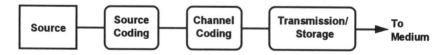

• Reception/Recovery/Decompression

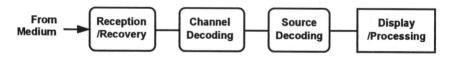

Figure 1.1 – Generic Model of a Digital Compression System

detail. Such a generic overview is shown in Figure 1.1. We will repeatedly refer back to this drawing in future installments of this column as we consider sundry aspects of ATV systems.

Figure 1.1 is a block diagram that shows elements of an ATV system without regard to where they are located physically or how they are connected together. Such details are among the types of things we will examine as we look at systems more closely in the future. In fact, a number of the blocks in our diagram may occur at several places in the overall system and may be concatenated.

Looking at each block briefly, the Source block includes all the functions of production and post production for use in the system. In an indication of how complex the system can get, the Source block may incorporate equipment that utilizes the other blocks for functions such as, for example, recording. In this way, the elements can become concatenated, and we can end up with a multiplicity of layers of the operations.

The Source Coding block is where all of the image compression takes place. It includes functions that sort out the information that must be sent to make a picture from the large amount of redundant information that is not required to be sent to make the picture. The Source Coding block is usually where the largest part of the system complexity is located.

The Channel Coding block takes the data derived in the source coding function and converts it to a form that can be sent through the medium to be used. Thus the Channel Coding may very well be different for terrestrial broadcast, cable, satellite, wireless cable, or tape recording, even though they carry the same form of source coding. Included in this function is any modulation required. The output signal usually has analog characteristics that embody the digital information.

The Transmission or Storage block takes the output of the appropriate channel coder and converts it to the final form for transmission or storage. For terrestrial broadcast, satellite transmission, or wireless cable, this is the transmitter and antenna. For cable, it is the headend processor and/or converter. For tape recording, it is the head drive amplifier. And for fiber optic transmission, it is the laser and its drivers.

The Reception or Recovery block captures the analog signal that carries the digital data and does the analog processing necessary to permit recovery of the digital information. Thus this block includes such functions as channel equalization and multipath elimination (ghost cancelling) to allow extraction of the data with the fewest errors possible.

The Channel Decode block performs the inverse of the channel coding operation, actually extracting the data from the received modulated signal.

The Source Decode block converts the compressed signal back to a form that can be displayed or further processed. This is the inverse of the source coding.

The Display or Processing block represents the ultimate user of the picture that has been sent through this particular combination of operations. Here the picture is viewed, processed for combination with other pictures, processed for further transmission, or the like. The output at this point may, in fact, be the input to a

Table 1.1 – High Definition Systems Under Consideration by ACATS

System Name	Proponent
Narrow-MUSE	NHK (Japan Broadcasting Corporation)
DigiCipher	General Instrument Corporation
Digital Spectrum Compatible HDTV	Zenith Electronics Corporation/AT&T
Advanced Digital HDTV	Advanced Television Research Consortium
Channel Compatible DigiCipher	MIT/General Instrument Corporation

similar process that conditions the picture for another transmission channel, another storage medium, or another purpose.

THE PLAYERS

There are a number of systems currently proposed for Advanced Television. They fall into the two categories of HDTV and Multichannel systems. Some of them have related entries in both categories using fundamentally the same system approach. All of them utilize pretty much the same underlying compression technologies, arranged in ways that their proponents feel will result in better performance. Some share common channel coding (modulation schemes). We will consider them all in future editions, as we look at the technologies they embrace.

The High Definition systems, all under evaluation by the FCC Advisory Committee and the Advanced Television Test Center, together with their proponents, are listed, in order of their testing, in Table 1.1.[1]

Table 1.2 – Multichannel Systems Proposed by Various Proponents

System Names/Descriptions	Proponent(s)
Proprietary DCT/4-VSB	AT&T/ComStream/News Datacom
DigiCable/DigiSat	General Instrument Corporation
MPEG-based/QAM(cable)/ QPSK(satellite)	Oak Communications/Leitch Video/ C-Cube
MPEG-based/QAM(cable)/ QPSK(satellite)	Philips Broadband/Philips Consumer/ Compression Labs/Hughes Network Systems
Vector Quantization/4-VSB	Scientific Atlanta
MPEG-based/QAM(cable)/ QPSK(satellite)	Thomson Consumer Electronics

---

[1] The last four systems and proponents in the table all eventually coalesced into the Grand Alliance. See Chapter 2 – "The Special Panel Meeting and Its Aftermath."

The multichannel systems are under evaluation in a number of places, foremost among which is the Cable Television Laboratories (CableLabs). The system names or the basic system techniques and their proponents (in alphabetic order) are listed in Table 1.2.[2]

COMING ATTRACTIONS

Future installments of this column will address the whole range of issues surrounding ATV and its technologies. We will not be limited to just the technologies but will look at such matters as the economics of different approaches, implementation choices, and the impact of regulations on system design. A list of a few of the items to be covered (in no particular order) includes:

Motion estimation/motion compensation
Discrete cosine transform
Sub-band coding
Huffman coding
Transmitter/antenna powers and sizes
Interleaving
Channel coding schemes
Channel equalizers and ghost cancelling
Production levels and performance
Production standards choices
Economics of ATV implementation
Timing of ATV implementation
System architectures
Networking
Pass-through operation
Post production of compressed signals
MPEG
Audio compression
Auxiliary data channels
Cellular operation possibilities

Because of its importance to the starting of planning for broadcast station implementation of ATV, next time we will take a look transmitters, transmission lines, and antennas and the power requirements that will be placed upon them.

---

[2] Many of the alliances between proponents have changed during the years since this was originally written. The number of players has increased over that time, and the combinations of organizations working together on various projects have multiplied. Some of the techniques that were listed have disappeared from consideration, and others, notably MPEG-2, as reflected throughout this book, have become prevalent.

2

The Special Panel Meeting and Its Aftermath

The Special Panel of the FCC Advisory Committee held its long-anticipated meeting February 8-11, [1993]. It was introduced by a number of participants in their opening remarks as an "historic" event. It wasn't – in the sense that it did not pick a single Advanced Television standard for the United States, the purpose for which it was originally scheduled. Nevertheless, it marked an historic transition in the development of Advanced Television – both for what happened during the sessions and for what was going on surrounding and caused by them.

There are three important things that happened the week of February 8. Two of them are now history, and one has the potential to lead to a very important change. This last one may be history by the time you read this (this is being written on February 12) or shortly thereafter. Because of the significance of these events, we will dispense this time with our usual examination of the technology and its implications. Instead we will try to put the happenings of the last week into perspective.

THE EXIT OF NHK AS A PROPONENT

The first important thing that happened at the Special Panel meeting was the elimination of the Narrow MUSE system from further consideration. This is not a surprising outcome; most participants and observers had expected it for a long time. But the Special Panel meeting was the first time that all of the results of the testing were examined by a body with the authority to make any real decisions. The weight

First published March, 1993. Number 9 in the series.

of the evidence showed that Narrow MUSE could not compete with the other systems, both in terms of the quality of the images delivered and in terms of the delivery of signals and the coverage possible. Thus the only remaining analog transmission system will not proceed to additional testing.

Even though Narrow MUSE is now gone from the process, it is important to recognize the contributions of NHK to the development of, first, High Definition Television (HDTV) and then Advanced Television (ATV). NHK (Nippon Hoso Kyokai – The Japan Broadcasting Corporation) invented HDTV. Starting in the early 1970's, under the leadership of Dr. Takashi Fujio, NHK engineers conceived the fundamentals of what we call HDTV today – double the horizontal and vertical resolution, for instance. Then they developed equipment to demonstrate the required techniques and to help in determining workable parameters. Eventually they settled on the 1125 lines and 60 Hz field rate of the system they currently support.

Next, the NHK engineering organization set about making the 1125/60 system practical. Working with manufacturers, but often designing the necessary components or circuits themselves, NHK engineers brought first cameras and monitors, later videotape recorders and signal processing equipment to the point that HDTV could be demonstrated to others. Their first public demonstrations in the United States were at the SMPTE Television Conference in San Francisco in February, 1981.

By 1984, NHK had developed a system for transmitting HDTV. It is called MUSE (for Multiple Sub-Nyquist Sampling Encoding) and was originally intended for satellite broadcasting in Japan. MUSE was demonstrated as an over-the-air terrestrial transmission system using two broadcast channels in early 1987. Later that year, when the FCC Advisory Committee on Advanced Television Service was established, NHK was requested by the National Association of Broadcasters to develop a version of MUSE for operation on a single 6 MHz channel. Narrow MUSE was the result.

Throughout the activities of the Advisory Committee, NHK and its representatives have acted as the models of gracious, collegial competitors. Starting with Yozo Ono and especially for the last several years with Keiichi Kubota, NHK has supported both the Advisory Committee process and the other proponents. Even at the moment their system was being set aside, at what was obviously a difficult time, Dr. Kubota retained his sense of humor in wishing the remaining proponents well. He also thanked the Advisory Committee for the fair treatment received by NHK and indicated his management's acceptance and support of the decision.

Of the many complimentary comments offered by members of the Special Panel at the point of reaching the decision, perhaps Joe Flaherty of CBS encapsulated the contributions of NHK best when he said, "I personally believe that other proponents would not have pursued their own concepts so aggressively had it not been for the leadership – and the pioneering leadership – of NHK. .... They, in fact, are at this moment still helping us with test materials we expect to use in a second round of testing. .... I think that we owe NHK a very large debt of gratitude, as a group of engineers, and a group of specialists in this field, for bringing us where we are today." None of us should ever forget NHK's contribution.

DIGITAL!

The second important thing that happened at or as a consequence of the Special Panel meeting was the confirmation that the future of television in this country, and probably throughout the world, will eventually be digital. This follows naturally from the removal of the last analog proposal from consideration. It also is a result long expected by most participants in and observers of the process. One has only to inspect earlier installments of this column to see the trend. Yet there was always previously a tentativity to this conclusion. Now it is final.

It will take time. Analog transmission will remain for fifteen or sixteen, maybe seventeen, years. Analog production will remain viable for a long time. But the day is coming when we will no longer think in terms of sync pulses but rather sync words. Line counts will be replaced with blocks, superblocks, and macroblocks. Headers and descriptors will give us the opportunity for tremendous flexibility in the material we send through systems and through the air. Compression will offer the possibility of scaling the amount of information we send or store about a picture to the requirements of particular applications. On Wednesday, February 10, 1993, digital television became a certainty.

FROM COMPETITION TO COOPERATION?

Much of the activity of the Special Panel meeting surrounded the filling in of a table that listed various performance attributes of ATV systems. This included items such as video performance on scene cuts, use of headers and descriptors, effective radiated power required, and interference from and to NTSC. There were 61 categories in total, and the proponents were awarded asterisks ("stars") if their systems showed a significant advantage over other systems in a category. The proponents politely argued back-and-forth among themselves and with members of the Special Panel as to why they should be awarded each of the stars – to the point that the table came to be called the "Star Wars chart."

In the end, each of the digital proponents accumulated, within +/- one star, the same number of stars as the others. In addition, the Special Panel, just as all of the Advisory Committee Working Parties that had tried before it, was unable to attach any value or weighting to any of the categories. Thus it became an impossibility for the Special Panel to differentiate and thereby select or de-select any of the remaining systems.

This is not to say there are no differences between systems. DigiCipher and AD-HDTV were clearly the best in image quality. DSC-HDTV and Channel Compatible DigiCipher were the best in transmission characteristics. But no one can say definitively that one of these is more important than the other.

All of this led to a recommendation by the Special Panel that further testing should be conducted, with each of the systems incorporating improvements that were approved by a Technical Subgroup of the Special Panel in November. The improvements are likely to make the systems only more difficult to differentiate in the future. And so on, and so on....

The message to the proponents from all this is clear: The competitive Advisory Committee process has been very good up to this point in pushing you to develop the best technology you could. In the future, the process will be less and less able to tell you apart. If you want to avoid years of meetings in the future like the one you just went through, find a way to settle this among yourselves outside the Advisory Committee process.

Indeed, if the metaphor of dating and courtship can be applied, the Special Panel meeting was a square dance – lots of noise and activity and fun, but no commitment. The real courtship dance, the intimate slow dance, was taking place elsewhere in Washington, where the CEOs of a couple of the major players and the decision-makers from some of the others were known to be meeting. An effort to arrive at a "grand consortium" was under way[3]. The problems of achieving such a grouping are no easier now than before. What is different now is the recognition that no proponent can win the contest on its own. With the analog systems now all out of the way, the digital systems can get together to forge a solution.

In trying to gather the best techniques from each of the digital systems into a single, optimum system, the proponents will face a major hurdle before they start: What scanning system should be used – interlaced or progressive? In fact, this issue was the focus of attention of the proponents in background negotiations taking place throughout the meetings. A special demonstration was arranged for the proponents at the Advanced Television Test Center (ATTC) starting at 9:30 am on Friday, February 12 – the morning after, as it were – to let them closely examine the two scanning methods.

---

[3] It actually came to be called the "HDTV Grand Alliance," or, simply, Grand Alliance – often referenced as the "GA." I got the concept right but missed on the name finally chosen.

Once a decision on the scanning technique is reached, an agreement to work together will have to be forged among the proponent organizations. From that point, the estimates I was given range between 10 months and 1 year for a testable hardware implementation to emerge.[4]

We should all wish the proponents well in their efforts to come together. If they fail to cohere, we face the possibility of years of indecision followed by years of legal battles. If it has not been published by the time you read this, hope for an announcement soon.[5]

---

[4] The GA hardware, in fact, moved into the laboratory at the Advanced Television Test Center (ATTC) on March 31, 1995, nearly two years after the Special Panel meeting. Bringing together the "best of the best," as the Grand Alliance did, was a more daunting task than had at first been recognized by the GA members. A decision was not reached on the last of the sub-system designs (the RF modem, after a "bake-off" between competing implementations) until early 1994, so, in reality, it took just a little over a year from that point to deliver testable hardware.

[5] It ultimately took until May, 1993, for an agreement to be reached between the four remaining proponents – under pressure from the imminent restarting of testing (due to begin the very day the matter was finally settled) and the world-class negotiating skills of Richard Wiley, chairman of the FCC Advisory Committee on Advanced Television Service (ACATS).

3

ATV System Recommended – What's Next?·

On October 31st, [1995,] the Technical Sub-Group of the FCC Advisory Committee's Special Panel met and approved the various reports and documents on the Grand Alliance Advanced Television system. They forwarded those materials to the blue ribbon Advisory Committee itself, with an endorsement that the Grand Alliance system be recommended to the FCC as the future television system for the United States. By the time you read this, the Advisory Committee will have met, on November 28th, and is expected to have approved the recommendation.

This is a major milestone in the effort to develop an Advanced Television system. It is one the Advisory Committee and all of its many participants have strived to achieve for eight years. But it is not the end of the process by any stretch of the imagination. Rather, it is almost the end of the beginning. Nevertheless, it provides a good opportunity to ask and try to answer the questions we will address this chapter: So, what comes next? And beyond that, what remains to be done?

NEXT STEPS — FCC

In some ways, the FCC has already begun to take its next steps with its release, in August, of the first of a group of three expected final Notices of Proposed Rulemaking (NPRMs) necessary to put the Advanced Television Service into place. That first final NPRM, with comments due November 15 [1995] and

---

First published December, 1995. Number 35 in the series.

reply comments due January 12 [1996], seeks essentially to complete the definition of what the ATV service will be. It addresses a wide range of questions about the structure of the ATV service, who will be licensed and under what circumstances, what comprises simulcasting, the lengths of the application/construction and transition periods, whether and how to place requirements on receiver capabilities, must carry and retransmission consent issues, and the like.

Many of these matters the Commission has addressed in previous NPRMs, but much has changed with the switch to digital technology subsequent to the FCC's last NPRM in 1992. Thus the Commission has taken the appropriate step of revisiting those prior decisions potentially impacted by such a major transformation in what ATV will be. At the same time, it has reaffirmed its belief in a number of important characteristics of the ATV service, such as the use of 6 MHz channels, while leaving the door open for comments. The Commission has also asked how best to recover the spectrum ultimately vacated by broadcasters and whether to require a second channel change by some stations, after the end of NTSC broadcasting, in order to "repack" the spectrum and free more for other uses. Upon completion of this particular rulemaking, the true nature of the ATV service should be relatively clear.

Yet to come from the FCC are two more NPRMs that will implement the ATV service. The first of these will adopt technical standards. It is expected to result directly from the recommendation to the Commission by the Advisory Committee. It should be based upon the standards documented by the Advanced Television Systems Committee (ATSC), which undertook to adopt the Grand Alliance system as a voluntary industry standard. The Commission will probably reference the ATSC documents without incorporating all of the details into the Rules. There may be a certain amount of the system definition that will be absorbed into the Rules, however, especially whatever is necessary to make it clear that the ATSC standards are the law of the land.

The Grand Alliance members would like to see this next NPRM issued in December, within days or weeks of adoption of the system recommendation by the Advisory Committee. My latest reading from Commission staffers, though, is that it will probably be early January before this Notice issues. There apparently is a lot of work to be done to turn the documentation to be submitted by the Advisory Committee into proposed Rules and to write the Notice. This will be stretched out by the various holidays that occur over the next month or so.

(By the way, if you would like to read the ATSC standards for yourself, you can download them from the ATSC's FTP site at `ftp.atsc.org` in the `/pub/standards/` directory. The files you want are: `a_53.zip`, the "ATSC Digital Television Standard;" `a_54.zip`, the "Guide to the Use of the ATSC Digital Television Standard;" and `a_52.zip`, the "Digital Audio Compression (AC-3) Standard." If you are new to this kind of documentation, start with the Guide. It is intended to be a tutorial. The documents are in Word for Windows 6.0c format, and you need the "c" release to read or print certain of the graphics. You can also download a Word viewer if you don't own a copy of the program. See the readme.txt file at the ATSC site for details.)

The last of the currently expected NPRMs will deal with the "three a's" of spectrum management: allocation, allotment, and assignment. Allocation, actually already determined for ATV, is the setting aside of a portion of the spectrum to be used for a particular purpose. Allotment is the provision of specific channels or frequencies to be used in particular regions or locales. Assignment is the matching of allotted channels with licensees in locations with allotments.

The broadcasters who filed a proposed channel pairing plan with the FCC earlier this year (see Chapter 23, "The Proposed Channel Pairing Plan," for more details) have requested that the Commission include the assignments in the same action in which it promulgates the allotments. The FCC, in the latest NPRM, discusses including only the assignment methodology. It remains to be seen which way this will come out. Work done in the Advisory Committee's Working Party on Transition Scenarios indicated that, if the FCC wants the quickest implementation, the broadcasters have the right idea in their suggestion of *a priori* assignments. But perhaps the Commission can devise something even better.

Assuming the FCC does not publish a channel pairing table along with the allotments in the third of the final NPRMs, it will have to turn the crank on whatever methodology it establishes to make the assignments. This will require another step, possibly initiated by the Commission, possibly initiated by individual broadcasters. In either case, the thinking of some FCC staffers seems to be that the Commission can be ready to accept applications as early as the third quarter of next year. If I were a betting man, I'd put my money on the fourth quarter, or the beginning of January, 1997.

NEXT STEPS — INDUSTRY

After the FCC completes its work on Rules, the industry's work will have really just begun. The Grand Alliance has been the focus of development efforts since mid-1993, shortly after the first round of testing. But the Grand Alliance ceases to exist once a recommendation has been made to and accepted by the Commission. The members of the GA end their cooperation and go back to being competitors. Yet the only part of the overall system that will have been developed and tested will be the transmission portion.

Still to be developed will be much of the technology for production, post production, distribution, and release facilities operating either in SDTV or in HDTV. The fundamental scanning standards or, more correctly, pixel matrices have been determined as part of the Advisory Committee process. Equipment only exists in the marketplace, however, for complete systems using the 480-line interlaced raster format (based, of course, on the 525 total lines of NTSC and equivalent component systems that have been in use for decades). Awaiting hardware availability are implementations of the 480-line progressive, 720-line progressive, and 1080-line interlaced raster formats.

Also to be accomplished, of course, is development of all of the consumer electronics equipment that will be needed if the Advanced Television Service is to succeed. Much of the design work for SDTV signals has already been done in other domains such as the set top receiver/decoders developed for cable, DBS, and wireless cable. Similar set top boxes will also be needed for broadcast ATV service, as we discussed in our last time together. [See Chapter 33, "Extracting SDTV Images from HDTV Streams."] None of the earlier work has handled HDTV signals, however, and appropriate decoder designs will have to be implemented to deliver HDTV to the larger displays that will support it, perhaps even demand it.

Aside from considerations of specific raster structures, there are entire categories of professional equipment that have not been offered for sale in forms appropriate for ATV use. Examples are tape recorders and other storage devices, switchers and digital effects equipment, graphics gear and character generators, monitors and test equipment, and the list goes on. At some levels, these have been awaiting the completion of standards and the development of a market. With completion and approval of certain of the necessary SMPTE standards during this past year, all related to raster-based, non-compressed signal forms, perhaps we will begin seeing some appropriate equipment in a few categories at next year's NAB convention.

DECISIONS, DECISIONS, DECISIONS

At other levels, however, there are still many decisions to be made by the industry as to how certain functions should be accomplished, and then standards must be developed to support those decisions. Included in this category are methods for switching compressed signals, the optimum bit rates to be used in distribution systems to permit seamless downstream switching and insertions for program continuity integration, the connectors and cables to be used for various categories and bit rates of signals, the types of signals to be used in transporting different forms of video, audio, and data within and between facilities, and so on.

Many of the required decisions will be based on the possibilities for new forms of operation. For instance, SMPTE and a number of manufacturers have been working to establish a new profile within the MPEG-2 matrix of profiles and levels enabling 4:2:2 operation at the main level (which encompasses several of the SDTV forms of operation). This 4:2:2 profile will enable MPEG-2 compression to be used in many studio situations, thereby permitting a common infrastructure to handle both studio and distribution signals equally easily.

With equipment becoming available for the new 4:2:2 profile fairly quickly (at NAB '96 is likely), the possibility then arises of carrying uncompressed, raster-based signals through the same infrastructure. If a packetized transport system, such as the MPEG-2 "systems" layer, is assumed for the compressed signals, why not use it for the raster-based signals, too? This would require forming the raster-based signals into packets — perhaps each line of the image could be carried in a packet. Then, both raster-based and compressed signals could be conveyed through the same paths, minimizing and possibly eventually eliminating the need for separate systems.

Probably the most difficult of the issues yet to be dealt with is the matter of switching of compressed signals. Because of its nature as a tool kit with many tools available to the encoder designer and/or operator and with only upper bounds on the many parameters that define a signal, MPEG-2 has the potential to be chaotic in its implementation by different users. This could easily lead to the inability to switch between signals that are compressed in different locations. To avoid this likelihood, the industry must agree on limiting the choices of parameters to certain specified values that are universally used. This will minimize, but not eliminate, the need for a multitude of copies of material at different bit rates to be created and inventoried so that they can be successfully integrated at facilities that have chosen different operating values.

(For a complete treatment of the switching problem, see my paper "Switching Facilities in MPEG-2: Necessary But Not Sufficient" in the December, 1995, issue of the *SMPTE Journal*, on pages 788-802. It explains the workings of MPEG-2 from a switching point of view, examines the difficulties of achieving clean switching between compressed streams, and offers suggested solutions to the problem. I have had this paper thoroughly checked out at places like Bell Labs, and I am told that it captures the issues correctly and delineates the most likely workable solutions. It is supporting discussions within a SMPTE Working Group set up for the purpose of handling switching and synchronization matters.)

Associated with the switching problem is the question of what data rates to use for distribution when downstream insertions are expected. There have been assertions by some that benefits might accrue from sending a higher bit rate to the final point of program continuity integration, thereby permitting higher quality to be attained after the decoding, switching, and re-encoding that will be necessary if other than cuts-only switching is implemented. If such a higher bit rate is used, decisions must be made as to how the additional bits should be allocated. If done properly, there is the potential to avoid complete decoding and to enable switching with effects at some intermediate level of compression.

INDUSTRY EFFORTS

There are two principal industry efforts that have been announced so far to deal with the matters that we have been discussing. The first is a private undertaking with government matching fund support through the Advanced Technology Program of the National Institute of Standards and Technology (NIST). It is led by the David Sarnoff Research Center (DSRC) and includes eight other companies. Its goals are to develop and make cost effective all of the technologies that are necessary to feed the GA transmission system. It will assemble an experimental ATV station at DSRC to serve as a test bed for integrating and testing the interoperability of the hardware produced by the various team members.

An open effort to deal with many of the issues we have been discussing is the establishment by SMPTE of a new Technology Committee on Packetized Television Technology. It will bring together in one place coordination of all of the on-going SMPTE work in areas such as systems design, switching and synchronization, interfaces, and compression issues. Five working groups that were split under two other committees have been brought together in the new committee. Other new working groups will be set up as additional work is identified. Ultimately, many of the standards and recommended practices that SMPTE has developed over the years will have to be modified or replicated to support packetized television.

In addition to the new DSRC and SMPTE efforts, the ATSC will continue its work. This will at least be in the form of supporting the new ATSC standards through the FCC rulemaking process and, presumably, through any changes that initial experience with implementation shows to be necessary. The continuation of the ATSC was in doubt for a while, but its Executive Committee recently decided unanimously to carry on. There is also interest within ATSC to move into new areas of work, and what they will be is currently being explored.

In the meantime, the ATSC will need a new chairman and probably a new executive director. Jim McKinney, the current chairman, will be leaving in mid-December, and Bob Hopkins, the executive director, is known to be looking for his next opportunity, both having completed what they went to the ATSC to do. The industry owes both a deep debt of gratitude for their masterful efforts in pulling together consensus among disparate industry segments and interests. The ATSC standards, mentioned earlier, along with many of the characteristics of the final ATV system design are tributes to their leadership of the ATSC's efforts.

The point that has been reached in the establishment of Advanced Television as the broadcast medium of the future is a most significant milestone. As we have seen, it changes the nature of the effort in many ways. At the same time, it assures that broadcast television will indeed fit over the long term into the broader context of ATV that we discuss in this column.

PART 2

Into the Woods

After over two years of writing about some of the details of Advanced Television technology in its many forms, it seemed time to expand our horizons and look at the big picture. In trying to write the first paragraph of such an overview, the old phrase about not being able to see the forest for the trees came to mind. Out of that grew a metaphor that tied together a four-part series in which we explored the status of applications, players, systems, and technology.

Although it originally came after a number of the articles that provided technical descriptions of various facets of Advanced Television technology, the series nevertheless makes a good place to start working our way into the technical side of things. It puts a lot of the technology into the context of the applications where it will be used and the industries that will use it. In the process, it provides references to many of the technical descriptions that now appear in later chapters in this book.

Among the most important material in this grouping is a fairly thorough explanation of the concepts of Profiles and Levels in MPEG-2 that appears in Chapters 4 and 5. Also included in Chapter 7 is a moderate amount on alternate distribution methods that you may find helpful and that does not appear anywhere else in this book.

Now let's see what we can do about "Seeing the Forest, Not Just the Trees." Please join me as we go "Into the Woods."

Seeing the Forest, Not Just the Trees

4

For somewhat over two years now, we've been looking at a few of the myriad aspects of the coming world of Advanced Television. Except for the first installment, our examination has studied the details, albeit at a rudimentary level, of the many technologies that will constitute the television of the future. With the upcoming SMPTE Technical Conference theme of "The Digital Era ... Ready or Not?," now seems a good time to step back and see what has been happening in the forest while we have been scrutinizing the trees. While we're at it, maybe we can see a little bit of who's been doing what to whom, as well.

This, too, is probably a good time to remind ourselves just what it is that we mean by the term "Advanced Television." Harking back to our first opus, we said, "Advanced Television, as we ... use the term here, is an agglomeration of techniques, based largely on digital signal processing and transmission, that permits far more program material to be carried through channels than existing analog systems can manage. In this sense, HDTV (High Definition Television) is a subset of ATV (Advanced Television). ATV is also a new way to look at the businesses in which we are involved — one that allows for a wide range of new or improved services impossible with earlier forms."

Note the inclusion of the word "businesses" in the preceding description. While the technology has developed rapidly over these past two years, some of the businesses in which it will be used have changed rather extraordinarily over the same period. It is probably just as instructive for our purposes to look at

---

First published September, 1994. Number 22 in the series.

those changes in the businesses that will apply the technology as it is to look at the technology itself. Here follows something of a traipse through the woods to take in the scenery for a change.

MAP OF THE FOREST

On a number of occasions in our looking at some segment or another of compression systems, we have used the diagram in Figure 4.1 to help us see just where the trees that we were inspecting were in the forest. It is a diagram of a generic digital compression system. Now, let us use Figure 4.1 to help us look at the forest itself and to find our way from clearing-to-clearing. (For those of you who are saving this series of articles – many of you have told me that you are – you can find a detailed description of Figure 4.1 in the first article in the series – July, 1992 [Chapter 1]; the 13th – September, 1993 [Chapter 15]; and the 19th – June, 1994 [Chapter 16]. The most complete description was the first one.)

We will work our way through the several blocks in Figure 4.1, considering both the encoding and decoding functions together. Remember that the decoding is essentially the inverse of the encoding process, although it normally has many different types of complications that make it quite a distinct function. All the while, we will see how these several stages fit into the various applications that expect to make use of them.

USERS AND APPLICATIONS

Before we look at the forest itself and wander into the clearings to see how beautiful or unsightly things are, we should learn who else is going to be traversing the same path we will take. This will help us to better understand the impact of what we see in the forest upon those who will be using the things to be found there. There are a large number of potential users and applications

- ## Compression/Transmission/Storage

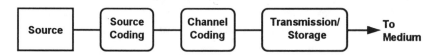

- ## Reception/Recovery/Decompression

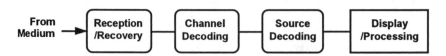

Figure 4.1 – Generic Model of a Digital Compression System

that will be depending upon advanced television for expanding their business opportunities in the future.

The users of advanced television will be entire industries as well as individual businesses. Among the first to apply this technology on a moderate scale are those supplying equipment for off-line, non-linear video editing, followed by suppliers of video servers for commercial insertion in cable systems. These are industrial applications with relatively small quantities involved. They have typically used MPEG-1 encoding, with some amount of wavelet encoding occuring more recently. Those using MPEG-1 are likely to upgrade to MPEG-2 when it is available.

The first wide scale applications, entailing consumer quantities, have been two satellite delivery systems – those of PrimeStar partners, using C-band direct transmission, and of DirecTV, using high power Ku-band direct broadcast satellite (DBS). PrimeStar started with a proprietary encoding system from General Instrument (DigiCipher I) and will be upgrading to another proprietary system that resembles MPEG-2 (DigiCipher II). DirecTV, on the other hand, is the first major application that is delivering decoders that are very close to MPEG-2 in their capabilities. Transmitting something closer to MPEG-1 at the start, it will soon move to almost full MPEG-2.

Cable operators and wireless cable operators are likely to be the next big users of advanced television techniques. There are currently over 3 million units ordered from an assortment of manufacturers. These are likely to be just the beginning, but the industry is waiting to see these delivered before committing to more. These generally embody MPEG-2 technology, and some also include proprietary capabilities close to those of MPEG-2 but with some different trade-offs.

The telephone companies have been saying for some time that they, too, expect to enter the video distribution business. They hope to use video delivery to support the cost of the advanced digital fiber optics networks they intend build in their territories. The expectation is that these systems will all use MPEG-2 encoding. The interesting part will be the form of the data networks they will build to transport the signals.

Finally, we come to broadcasters. One of the largest development efforts for advanced television has been devoted to making it work for over-the-air transmission on the frequencies and in the channel bandwidths of broadcast television. It is this work that stimulated much of the other development that now or will soon support the other mass distribution media. All of the other media, however, are working toward implementation of standard definition television, probably with some widescreen enhancement capability, as their first order of business. Most consider HDTV, if at all, as an afterthought.

But development for broadcasting started out as being exclusively for HDTV. The FCC proceedings have so far assumed that the broadcast form of advanced television will be HDTV. Recently, however, under pressure from broadcasters, there has been some movement in the FCC Advisory Committee and in Congress toward allowing broadcasters the flexibility to choose between HDTV and multiple SDTV signals, or some combination, for transmission. While this is far more a political choice than a technical one, involving intense lobbying and power struggles between industry segments, the technology to permit any permitted outcome is already in place.

SOURCES

Sources for advanced television systems can range from standard definition television (SDTV), through all kinds of film formats, to high definition television (HDTV). Film will always have to be transferred first to video before further processing. An important issue for such film transfers in the future will be the frame rate at which they are recorded.

Currently, film transfers are recorded either at 60 fields per second using 3:2 pull-down or at 50 fields per second with a 4 per cent speed error. Virtually all compression systems remove the 3:2 pull-down in 60 Hz versions as a means of reducing redundancy in the transmitted information, reinserting the 3:2 relationship in the display when appropriate. This processing could be simplified in the future by simply recording the original transfers at 48 interlaced fields or 24 progressive frames per second. Such an approach is currently under discussion within the production and post production communities.

SDTV signals can accommodate both standard 4:3 and widescreen (16:9) aspect ratios either by mapping the pixels differently or by changing the number of pixels in a line to switch between the two. HDTV, on the other hand, is essentially a 16:9 medium; material originated in other aspect ratios is handled by placing borders on the top or the sides of the image, depending on what is required to map any particular source to 16:9.

There has been some recent controversy over the choice of 16:9 as the widescreen aspect ratio for advanced television, with strong unhappiness expressed with this value by cinematographers as evidenced in the pages of *TV Technology* for the past couple of issues. Considering the worldwide adoption of this value, its general acceptance for more than a decade, and the tens of millions of dollars of investment in tooling for 16:9 by the world's manufacturers, there is virtually no chance this value will change, despite the cinematographers protests.

The coming of digital video compression will accelerate the move in the studio to component operation. Noise in the image to be coded quickly reduces the quality of the compression encoder output because the encoder cannot tell that the noise is not moving image information, and it tries to encode it. This leaves less data capacity for carrying the real picture information.

Residual subcarrier left in the signal after composite decoding appears just like noise to the compression encoder. Since even the best composite decoders generally leave some subcarrier residue, the compression encoder will almost always produce a lower quality image from a previously composite-encoded then decoded signal than it will from one that has never been encoded, i.e. a component signal.

The fact that the best composite decoders require frame stores and are thus somewhat expensive exacerbates the problem since they cannot be afforded ubiquitously. All this leads to the conclusion that component operation is the best choice for studio systems that will feed digital video compression encoders. This conclusion is underlined by the fact that virtually all of the compression systems are themselves inherently component in nature, so sources will consequently be fed to them in their native form.

VIDEO SOURCE CODING

There are a number of divergent forms of compression that can be applied at the source coding stage. The objective of each, of course, is to eliminate redundancies in the image and to store or transmit only the minimum information essential to allow the decoder to reconstruct the original. These break down into three general categories of coding techniques: (1) transforms that concentrate the image energy into a smaller number of coefficients than would otherwise be required, (2) algorithms that describe images through a series of formulas, and (3) motion estimation and compensation processes that reduce the burden on the transforms that follow them.

Categories 1 and 2 are competitive techniques that are represented by the discrete cosine transform (DCT) and fractal encoding respectively. Sub-band coding and wavelet coding are additional types of transforms in category 1. Motion estimation/motion compensation is normally applied in conjunction with a transform such as the DCT.

The most widely known and used scheme is that embodied in the internationally developed standards of the Moving Pictures Expert Group (MPEG), a jointly formed body of the International Telecommunications Union (ITU) and the International Standards Organization (ISO) – two United Nations treaty organizations. The MPEG scheme used in the first two MPEG standards, MPEG-1 and MPEG-2, combines motion estimation/compensation and the DCT along with variable length coding in a manner that provides a great deal of flexibility in their application. (We covered the operation of MPEG-style systems at some length in previously-written columns – see Part 3 for a five-part treatment of the subject.)

MPEG-1 was originally developed as a means for compressing small (i.e. low pixel count, hence low resolution) images primarily for storage on CD-ROMs and for display on computers. It was adapted for other applications (such as the ATRC submission of AD-HDTV to the FCC Advisory Committee) because it was all there was at the time that was in the process of being standardized. MPEG-1 and its variants and proprietary upgrades has served as the foundation of many of the operating compression systems because of its early adoption and the availability of chips for its implementation.

MPEG-2 was developed specifically for television applications of all types ranging from videoconferencing up through HDTV. It encompasses a very wide extent of source and display resolutions, storage and transmission data rates, and processing technologies. MPEG-2 introduced the concept of a packetized structure for storing and transporting the information. (See Chapters 16, 17, & 18 in Part 5 for an extensive three-part treatment of packetized digital video.) MPEG-2 provides a rich set of features and parameters that are described by a language (a "syntax") that carries the instructions from encoder to decoder as to how to reconstruct the image.

Another novel concept introduced with the development of MPEG-2 is a matrix structure of "levels" and "profiles" that put bounds on what a decoder conforming to the standard is expected to be able do. Because of the extremely broad spectrum of applications expected to be served by MPEG-2, this is a

Table 4.1 – MPEG-2 Profiles and Levels

Level	Simple	Main	SNR	Spatial		High	
High		1920 1152 60				1920 1152 60	960 576 30
High-1440		1440 1152 60		1440 1152 60	720 576 30	1440 1152 60	720 576 30
Main	720 576 30	720 576 30	720 576 30			720 576 30	352 288 30
Low		352 288 30	352 288 30				

Notes: 1. Matrix intersections with values shown are the only ones defined.

2. Values given in cells represent, from top to bottom, maximum samples per line, maximum lines per frame, and maximum frames per second.

3. Matrix intersections having two cells with values are defined as having two spatial resolution layers – an "Enhancement" layer, shown on the left, and a "Lower" layer, shown on the right.

4. The "SNR" and "Spatial" profiles allow for scaled coding in which the scaling occurs in the dimension indicated by the profile name.

necessary device for making it economical to implement. It effectively breaks up the ranges of possible parameters into chunks for which decoders can be advantageously optimized. Without such an approach, all decoders would be expected to handle the entire gamut of possible values included in the syntax. This might require, for instance, that a decoder intended for videoconferencing be able to handle HDTV signals. Clearly, this would be wasteful, and implementation would not happen this way. The latest version of the table of levels and profiles is shown in Table 4.1.[6] We will look at this in more detail next time.

There are significant differences between MPEG-1 and MPEG-2 that derive from their different intended applications and from the fact that MPEG-2 was designed later, when the technology was more mature and the problems to be addressed were better understood. To help in comparing and contrasting MPEG-1 and MPEG-2, Table 4.2 shows some of their principal characteristics with the areas of difference shaded.

We have been concentrating on MPEG video compression schemes to this point (and, indeed, in earlier outings as well) because it has had the largest investment of resources and is anticipated to be the most widely applied by far. It has become essentially the *de facto* standard in people's expectations even before it is a *de jure* standard. That does not, however, necessarily make it the best technique.

Your earnest writer has witnessed demonstrations of alternative video compression technologies that achieved image fidelity levels surpassing anything seen by him so far using MPEG techniques. These alternative methods have used various forms of wavelet transforms and fractal implementations. They sometimes have other benefits such as easier processing requirements. But the likelihood is that they will not achieve the very wide implementation expected especially for MPEG-2. This is because MPEG-2 has already reached critical mass with respect to high volume consumer applications. The alternatives likely will see use, but in situations where there is low concern regarding interoperability of systems and where they have particular benefits to offer such as high processing efficiency.

AUDIO SOURCE CODING

It is important to note that both MPEG-1 and MPEG-2 include audio compression along with the video upon which we have largely concentrated in this column over the past two-plus years. The original choice of audio

---

[6] A new "4:2:2" profile was added following publication of the original article. It essentially extends the Main Profile at Main Level in several respects to provide for 4:2:2 operation and higher bit rates. Some details appear in a table following Chapter 5. It was approved in January, 1996.

Table 4.2 – Comparison of MPEG-1 and MPEG-2 (Differences highlighted).

Characteristic	MPEG-1	MPEG-2
Coding method	Motion estimation, motion compensation, transform coding of displaced frame differences	Motion estimation, motion compensation, transform coding of displaced frame differences
Transform	Discrete Cosine Transform (DCT) w/8X8 blocks	Discrete Cosine Transform (DCT) w/8x8 blocks
Frame structure	Group-of -Pictures with Intra-coded frames, Predicted frames, and Bi-directionally predicted frames	Group-of -Pictures with Intra-coded frames, Predicted frames, and Bi-directionally predicted frames
Field/frame processing	Frame only (Progressive scan design)	Field & frame (Both Progressive scan & Interlace handled)
Variable length coding	Huffman with primitive coding tables	Huffman with refined coding tables
Systems layer	None provided	Packetized transport with provisions for flexibility, extensibility, and interoperability – supports private data transmission
Audio compression schemes	Musicam I (mono or stereo only)	Musicam I, Musicam 5.1, Dolby AC-3 (mono, stereo, or surround plus subwoofer)
Lip sync method	None provided	Program and channel time stamps used to control relative decoder timing
Error correction coding	None – must be provided as part of transmission/storage channel-coding system	None – must be provided as part of transmission/storage channel coding system
Access control	Not supported – must be provided as part of transmission/storage channel-coding system	Supported through Systems layer control functions – actual access control method not defined

compression method was the Musicam system developed by a consortium that included Philips, Matsushita, and other organizations. It was selected competitively from among a number of systems that were developed by several consortia for the MPEG-1 standardization effort.

Subsequent to the original selection of Musicam, two things have happened: (1) There has been acceptance of a 5.1 channel approach for surround sound in high end applications (the 0.1 channel is for a subwoofer), and (2) a number of new competing systems have been put forward that so far have tested better than Musicam in subjective evaluations. As a result of item 1, the original Musicam is being updated to Musicam 5.1.

As a result of item 2, an issue has arisen in the MPEG-2 standardization process over whether it is necessary for MPEG-2 audio to be backwardly compatible with MPEG-1 audio. This has thrown the MPEG-2 process into something of a tizzy with regard to the choice of backwardly-compatible (BC) or not-backwardly-compatible (NBC – not to be confused with the network) audio. There is a possibility that both forms will be allowed in the final documentation.

Some applications that have had to move forward without waiting for the MPEG-2 decision have been forced to decisions between Musicam and Dolby AC-3, the only other currently available contender. Examples of these choices are DirecTV (which uses Musicam) and the Grand Alliance HDTV system (which will use Dolby AC-3). We will have a complete treatment of digital audio compression systems, the on-going selection process, and other audio-related matters in an upcoming column.

CHANNEL CODING

Channel coding is where the applications really begin to diverge and where the differences between industries become very apparent. We will pick up with channel coding, right after looking at levels and profiles in more detail, when we continue next time.

5

Of Profiles, Levels, and Forests

In our last outing, we started a little jaunt through the woods of advanced television with the intention of taking in the forest for a change instead of examining the trees. We took a quick look at the map before beginning our treck; then we stopped at the clearings for users and applications, sources, video source coding, and audio source coding. This time, we'll continue into channel coding, where there are many forks in the road and where differences between industries can really get things going.

We ran into the concepts of profiles and levels when we looked at MPEG-2 source coding last chapter, and we promised to look at them a bit more this time. So before moving on to channel coding, we'll revisit profiles and levels to see how they help to organize MPEG-2 – a forest in itself – into groves of manageable proportions.

PROFILES AND LEVELS

As we've seen before, MPEG-2 is expected to cover a very wide spectrum of applications. On the low end, it is useful for low level videoconferencing and image database storage and retrieval. In the middle, it is intended for all kinds of entertainment distribution including broadcasting, satellite, cable, telco, and packaged media (digital videotape and digital videodisk). At the high end, it covers applications of high definition television, some of which are not even achievable on a practical basis with current production technology.

First published October, 1994. Number 23 in the series.

If MPEG-2 were a unitary standard, each encoder and decoder would have to process the signals for the entire range of applications. This would, for example, burden a videoconferencing system with the capability to handle very high definition images. The cost for doing this would make use of MPEG-2 unworkable for the videoconferencing operation. This is probably one of the more extreme, but many such examples can be cited.

To deal with this issue and a number of others, its developers chose to make MPEG-2 more of a toolkit than a fully defined standard. This allows an encoder to send an image to a decoder in many different ways. What is defined is the language that the decoder must understand, the words in that language, and the way sentences must be put together from those words (the "syntax"). But the encoder can select from the vocabulary and can construct sentences to describe the image in any way it chooses, so long as it follows the syntax.

Table 5.1 – MPEG-2 Profiles and Levels

Level	Profile						
	Simple	Main	SNR	Spatial		High	
High		1920 1152 60				1920 1152 60	960 576 30
High-1440		1440 1152 60		1440 1152 60	720 576 30	1440 1152 60	720 576 30
Main	720 576 30	720 576 30	720 576 30			720 576 30	352 288 30
Low		352 288 30	352 288 30				

Notes: 1. Matrix intersections with values shown are the only ones defined.

2. Values given in cells represent, from top to bottom, maximum samples per line, maximum lines per frame, and maximum frames per second.

3. Matrix intersections having two cells with values are defined as having two spatial resolution layers – an "Enhancement" layer, shown on the left, and a "Lower" layer, shown on the right.

4. The "SNR" and "Spatial" profiles allow for scaled coding in which the scaling occurs in the dimension indicated by the profile name.

The use of a toolkit allows different implementations of encoders and decoders that can be optimized for particular purposes and whose designs can improve with the technology while still remaining interoperable with one another. But it doesn't, by itself, solve the problem of equipment having to cover the entire, very broad range of applications. Once the possibility is opened up to the variability provided by a toolkit, however, it becomes possible to put bounds around the capabilities of specific encoders and decoders, limiting their uses to distinct classes of applications while still permitting them to conform to the "standard."

This means that not every encoder and not every decoder will work for every application. Nor will every decoder be capable of decoding the signals from every encoder. Thus encoders and decoders must be matched to one another. Matching such devices on a detail-by-detail basis would be a grueling task. In order to provide for this in a manageable manner, an array of standard subsets of the overall syntax has been created with the two dimensions of the array labelled as "profiles" and "levels." The array of profiles and levels is shown in Table 5.1, which is a repeat of last chapter's Table 4.1.

Devices operating within a particular intersection of the matrix must meet the defined requirements for that region of the standard. In this way interoperability of encoders and decoders can be assured. The matrix also serves as a shorthand way to describe the capabilities of any given MPEG-2 device. Thus an encoder or decoder can be described as conforming to the main profile at main level (alternatively, MP@ML or main/main), the high profile at high level (HP@HL or high/high), or the simple profile at main level (SP@ML or simple/main) to give just a few examples. (The full list of abbreviations for profiles and levels is given in Table 5.2.) Within each of these categories,

Table 5.2 – Abbreviations for Profile and Level Names

Profile	Profile Abbreviation	Level	Level Abbreviation
Simple	SP	Low	LL
Main	MP	Main	ML
SNR Scalable	SNR	High-1440	H-14
Spatially Scalable	Spatial	High	HL
High	HP		
4:2:2	4:2:2		

Note: The 4:2:2 Profile was added following original publication of this article. For details, see the section "4:2:2 Profile Update" at the end of this chapter.

Table 5.3 – Syntactic Constraints on Profiles

Syntactic Element	Profile				
	Simple	Main	SNR	Spatial	High
Chroma Format	4:2:0	4:2:0	4:2:0	4:2:0	4:2:2 or 4:2:0
Picture Coding Type	I,P	I,P,B	I,P,B	I,P,B	I,P,B
Scalable Mode			SNR	SNR or Spatial	SNR or Spatial
Intra-Coded DC Precision	8,9,10	8,9,10	8,9,10	8,9,10	8,9,10,11

Note: 1. Not all constraints on profiles are shown. This is a selection for
        sake of example only.

encoders and decoders can be expected to interoperate properly with one
another. Nonetheless, this does not preclude an encoder or a decoder from
having the capability to operate in more than one region of the matrix.

The differentiations between the several profiles and levels are expressed in
terms of "constraints" placed on various elements of the syntax, limiting the
use of certain features of the system or placing bounds on the values that can be
adopted by certain parameters. This can be seen in Table 5.1, where the values
given in each allowed array point are the upper bounds for sampling density in
the horizontal, vertical, and temporal dimensions, reading vertically within
each cell. Note that not all of the array points are defined; only the ones thought
to be of significant use have been documented.

Table 5.4 – Parameter Constraints for Levels

Syntactic Element	Level			
	Low	Main	High-1440	High
Vertical Vector Range – Frame Picture	-64:63.5	-128:127.5	-128:127.5	-128:127.5
Vertical Vector Range – Field Picture	-32:31.5	-64:63.5	-64:63.5	-64:63.5

Notes: 1. Not all constraints on levels are shown. This is a selection for sake
         of example only.

2. Values are in pixels.

There are many other constraints applied to the different profiles and levels. Some apply to all levels of a given profile; some apply to all profiles at a given level. Others vary in both dimensions of the matrix. Some further examples are syntactic constraints on chroma format, picture coding type, scalable mode, and intra-coded dc precision along the profile axis (as shown in Table 5.3); parameter constraints on motion vector ranges along the level axis (shown in Table 5.4); and upper bounds on luminance sample rates and compressed bit rates, and minimum requirements for video buffer verifier size (VBV – a mechanism for tracking the buffer fullness measure that we have described in previous columns – see Part 3 and, in particular, Chapter 12) that are defined on a matrix basis (shown in Tables 5.5, 5.6, and 5.7, respectively).

SCALABILITY

In addition to providing for a narrowing of the capabilities required of encoders and decoders, profiles and levels, as is apparent in some of the tables, also provide for several forms of scalability in MPEG-2 systems. Scalability uses multiple bitstreams to permit two different levels of image performance to be obtained without requiring independent transmission of the full information for each of the two performance levels (called "simulcasting" in MPEG-2 parlance — different from the FCC's use of the term). The underlying technique in each case is to send a base layer, called the "lower layer," using one of several methods, accompanied by additional information that can be processed so that, when it is added to the decoded lower layer, it forms a higher performance image, called the "enhancement layer."

Scalability can be beneficial for one of several reasons, and the choice of technique will depend upon the results sought. It may be desirable to permit lower cost decoders to extract a moderate performance image while higher cost decoders can extract a high performance image from the same MPEG-2 stream without duplicating a substantial portion of the data. This implies spatial scalability and might be used, for instance, to send a standard definition (SDTV) version of a program embedded in a high definition (HDTV) version.

Spatially scalable layers are indicated in Table 5.1 by listing them in side-by-side cells at a matrix intersection. In these cases, the enhancement layer is on the left, and the lower layer is on the right. Note the 2:1 relationships between the maximum values of each of the layer pairs. This means that, if both are operated at their maximums, the enhancement layer will have 8 times as much information as the lower layer.

Scalability may be desirable to transmit some of the data with high error correction overhead or to give it a high priority in a switching network to guarantee delivery at least of a minimum level of performance while data to make a higher performance image is sent with lower overhead or a lower priority. This implies SNR (signal-to-noise ratio) scalability and could be used

Table 5.5 – Upper Bounds for Luminance Sample Rate

Level	Spatial Resolution Layer	Profile				
		Simple	Main	SNR	Spatial	High
High	Enhancement		62,668,800			62,668,800 (4:2:2) / 83,558,400 (4:2:0)
	Lower					14,745,600 (4:2:2) / 19,660,800 (4:2:0)
High-1440	Enhancement		47,001,600		47,001,600	47,001,600 (4:2:2) / 62,668,800 (4:2:0)
	Lower				10,368,000	11,059,200 (4:2:2) / 14,745,600 (4:2:0)
Main	Enhancement	10,368,000	10,368,000	10,368,000		11,059,200 (4:2:2) / 14,745,600 (4:2:0)
	Lower					–
Low	Enhancement		3,041,280	3,041,280		3,041,280 (4:2:0)
	Lower					

Notes: 1. In the case of single layer or SNR scaled coding, the limits specified by 'Enhancement layer' apply.
2. Values are in samples per second.
3. The luminance sample rate is the product of samples per line, lines per frame, and frames per second.

Table 5.6 – Upper Bounds for Bit Rates

Level	Profile				
	Simple	Main	SNR	Spatial	High
High		80			100 all layers 80 middle + base layer 25 base layer
High-1440		60		60 all layers 40 middle + base layers 15 base layer	80 all layers 60 middle + base layers 20 base layer
Main	15	15	15 both layers 10 base layer		20 all layers 15 middle + base layers 4 base layer
Low		4	4 both layers 3 base layer		

Notes: 1. Values are in Megabits per second (Mb/s). 1 Mbit = 1,000,000 bits.

2. This table defines the maximum coded data rate for fixed bit rate operation and the maximum elementary stream rate for variable rate operation.

3. This table defines the maximum permissible data rate for all layers up to and including the stated layer. For multi-layer coding applications, the data rate apportioned between layers is constrained only by the maximum rate permitted for a given layer as stated in this table.

for transmission over networks such as ATM that are subject to congestion or over radio frequencies that are subject to fading and other disturbances.

Scalability may also be desirable to permit migration to higher levels of performance as the technology matures, allowing older decoders to continue to operate at their levels of capability while newer decoders yield improved imaging. This implies temporal scalability and might facilitate a move from interlaced to progressively scanned HDTV or from lower to higher frame rates. Although listed separately, some of these techniques can be applied jointly (as

Table 5.7 – Video Buffer Verifier Minimum Size Requirements

Level	Layer	Profile				
		Simple	Main	SNR	Spatial	High
High	Enh. 2					12,222,464
	Enh. 1					9,781,248
	Base		9,781,248			3,047,424
High-1440	Enh. 2				7,340,032	9,781,248
	Enh. 1				4,882,432	7,340,032
	Base		7,340,032		1,835,008	2,441,216
Main	Enh. 2					2,441,216
	Enh. 1			1,835,008		1,835,008
	Base	1,835,008	1,835,008	1,212,416		475,136
Low	Enh. 2					
	Enh. 1			475,136		
	Base		475,136	360,448		

Notes: 1. Values are in bits.

2. Buffer size is calculated to be proportional t the maximum allowable bit rate rounded down to the nearest multiple of 16 x 1024 bits. The reference value for scaling is the Main profile, Main level buffer size.

3. This table defines the total decoder buffer size required to decode all layers up to and including the stated layer. For multi-layer coding applications, the allocation of buffer memory between layers is constrained only by the maximum size permitted for a given layer as stated in this table.

can be seen in Table 5.3), resulting in multiple, simultaneous levels of scalability.

CHANNEL CODING

Returning now from our little side trip into the profiles and levels groves, we continue through the forest down the path shown on our map – last chapter's Figure 4.1. We soon come to the next clearing – channel coding. Channel coding is the section of the system in which compressed images and sounds, along with any other accompanying data, are formatted for transmission and/or storage. The Channel Coding block takes the data developed in the source coding function and converts it to a form that can be sent through the medium to be used. Thus the Channel Coding may very well be different for terrestrial broadcast, cable, satellite, wireless cable, or tape recording, even though they carry the same form of source coding. Included in this function is any modulation required. The output signal usually has analog characteristics that embody the digital information.

There are normally several steps in the channel coding process. First comes formatting into the final data stream, usually in the form of packets. Then comes addition of any error correction coding (ECC). Finally comes modulation of some type or another. The channel coding may be changed several times as a program or element flows through the production and distribution food chains.

For example, a particular program might be recorded on tape or disk using one channel coding; then be transmitted over a satellite using another channel coding; next be recorded on another tape or disk, possibly using the first channel coding again; and finally be transmitted over a broadcast or cable system using yet another channel coding. Along the way, the program is likely to be scrambled and unscrambled and to have conditional access control information added and deleted as part of the channel coding process. All of this points out that the channel coding must be able to be freely added and deleted as required by the system configuration at any particular time.

As we pointed out last time, MPEG-2 is likely to be the predominant encoding method for most entertainment and other electronic media distribution and communication. It specifies the packetization schemes to be used with its video, audio, and other (called "private") data. Two choices are offered: Program Stream packetization, in which variable length packets are used in their "native" form, and Transport Stream packetization, in which fixed length packets are used.

The choice of packet form depends upon the application — Program Stream packets are best for low error rate environments such as recorded media; Transport Stream packets are best for higher error rate environments such as radio frequency transmission. The fixed length of the Transport Stream packets makes it relatively easy to add ECC at an overhead rate appropriate to the transmission channel. (For a full treatment of MPEG-2 packetization, see Chapters 16-18 in Part 5.)

Because of MPEG-2's anticipated predominance, within the two categories, the packetization of systems will be essentially the same from one to another since MPEG-2 defines all of this as part of its "systems" layer. But this is where MPEG-2 stops. Error correction coding and anything beyond it are not included in the standard, and implementers are free to do whatever they feel to be appropriate. To some extent, this is necessary because it would be impossible for the MPEG committee to examine every application and coalesce around a single implementation for it. It does, however, lead to a state of competition and often confusion that continues through the transmission/storage domain.

The closest thing to a standard for ECC is the approach taken by the Grand Alliance (GA) for its proposed broadcast HDTV system. The GA appends 20 Reed-Solomon parity bytes to each 188-byte MPEG-2 Transport Stream packet to form a 208-byte block that is the basic unit transmitted. The particular R-S code chosen permits up to 10 byte errors per block to be corrected. Reed-Solomon codes are especially good at correcting impulse noise or burst errors, although they do not help much with errors from the white noise that interferes with the signal when operating near carrier-to-noise (C/N) threshold. Additional error correction techniques are usually used to help with white noise, and we'll discuss this more shortly.

Many other error correction choices can be made aside from those of the Grand Alliance. Each system designer can select the types and values most helpful to a given application. With no standardization, however, this means that channel encoders will have to be carefully matched to the channel decoders installed for each purpose. Very often, it may mean that the channel encoders and decoders may have to be purchased from the same supplier or that alternate suppliers will have to license techniques from a competitor.

This is the case, so far, in the realm of set top boxes for cable, wireless cable, and telco applications. Certain manufacturers have taken the lead in developing complete systems while others have either developed their own or licensed technology from the ones who have already done the development work. This leads to a large number of incompatible products in the marketplace. Although coalescence on certain approaches can be expected, if system implementers obtain equipment from different sources, they must be very cautious during the early days to ascertain that their equipment is well-matched from the transmission to the reception end of a channel.

Such issues continue as we progress through the remainder of the channel coding process. There are numerous possible choices for interleaving, which normally follows Reed-Solomon-type ECC coding and helps to spread the errors around, making them even less susceptible to burst errors. This is often followed by some form of additional coding that helps with bit errors caused by white noise, effectively extending the reception threshold to lower levels. This further coding is often of a type called "trellis" coding. At the receiver, it improves recovery of the signal in a noisy channel; hence it is often described as providing "coding gain." In trellis coding and similar techniques, the coding gain comes from sending more bits as overhead and using them for a form of error correction directly in the demodulator. Again, many designs are possible.

MODULATION METHODS

Trellis coding is intimately related to channel modulation. There are four major forms of modulation currently under consideration for various applications: PSK, QAM, VSB, and COFDM. Each of them offers several-to-many possible parameter choices. VSB (Vestigial SideBand) has been selected for the GA system. It uses 8-level or 16-level coding for broadcast and cable applications, respectively. 8-VSB provides a payload data rate of 19.3 Megabits per second (Mb/s), while 16-VSB provides double that – 38.6 Mb/s. In going from 8-VSB to 16-VSB, the payload data rate doubles even though the number of bits per symbol only goes from 3 to 4. This is because there is no trellis coding used for 16-VSB and, consequently, no trade-off of data rate for coding gain. (See Chapter 22 for a full treatment of VSB.)

QAM (Quadrature Amplitude Modulation) is probably the front-runner as the choice for virtually all applications other than broadcast and satellite. QAM comes in varieties with 16, 32, 64, and 256 points in their constellations. These represent 4, 5, 6, and 8 bits per symbol and can carry on the order of 16, 21, 27, and 37 Mb/s in a 6 MHz channel. It can also be used in much narrower channels at lower data rates. This can allow single-channel-per-carrier (SCPC) operation through satellites or video transmission over copper pairs. Exactly what all these numbers are depends on choices made for error correction overheads, use of trellis coding, and the like.

PSK (Phase Shift Keying) is similar to QAM with the amplitude component removed. It provides a constellation of points that describe a circle around zero carrier at the center. Thus the points have equal amplitudes and different phases. Most commonly used is QPSK, which has four points in quadrature around the center. Some systems use 8-PSK (8 points in the constellation), although it is less common. PSK systems are used for satellite transmission because they permit saturation of the transponder and result in the lowest C/N requirements at the receiver. The trade-off is that they require more bandwidth than the denser, amplitude modulation schemes.

COFDM is a technique that uses a multiplicity of PSK or QAM signals spread across a channel with the data to be sent divided between them. This is the method used in the new high speed data modems for personal computers. It is claimed to offer advantages for channels with multipath and other aberrations and may be particularly useful for distributed transmission systems where multiple transmitters in relative proximity to one another carry the same signals. (See Part 8 for four chapters devoted to COFDM.)

COFDM using multiple QPSK carriers has been developed for digital audio broadcasting (DAB) in Europe. There have been a number of proposals for using COFDM for digital terrestrial television broadcasting (DTTB) including HDTV, largely from Europe, but a group led by the NAB is also pursuing development of the technique as an alternative for use with the Grand Alliance system. In addition, a version of COFDM has been proposed for the ADSL (Asymmetric Digital Subscriber Line) system that some telcos are considering using to deliver video dial tone to their customers over copper pairs.

TRANSMISSION/STORAGE

Transmission and storage, in some senses, are where the rubber meets the road. They are where all of the technology we've been describing gets applied to store and deliver programs. We'll complete our forest walk next time with an overview of some of the applications and networks under development by cable operators, telcos, satellite broadcasters, terrestrial broadcasters, and others.

We'll include such schemes as hybrid fiber/coax, ADSL, video servers, and others.

4:2:2 PROFILE UPDATE

Just prior to preparation of this book, in January, 1996, the MPEG committee approved an additional profile intended essentially for professional use. It is named the "4:2:2 Profile" in order to describe its fundamental difference from the Main Profile at Main Level. There are other significant differences such as a much higher permitted bit rate and a maximum line count high enough to permit carrying much of the vertical interval.

Testing of various implementations of the profile has shown that appropriate choices of bit rates, group of pictures (GOP) structure, and other characteristics can permit multi-generation encoding and decoding followed by many types of post production processing and editing. Even chroma keying or color matting is possible under the right circumstances. While not intended for the very highest quality applications, it is likely to permit many professional applications hitherto not possible.

The name Professional Profile was specifically not applied to the new profile to reserve its use for a possible future, more capable and complete profile that might address even the highest quality applications as well as HDTV levels of performance. The fundamental characteristics of the 4:2:2 profile are shown in Table 5.8, in which are given the values that would have been included in all the previous tables in this chapter had they been completely reformulated. The pressures of meeting a publishing deadline, alas, prevent such a complete inclusion.

Table 5.8 – 4:2:2 Profile Fundamental Characteristics

Max. Samples/Line	720	Vertical Vector Range–Frame Picture	-128:127.5
Max. Lines/Frame	608	Vertical Vector Range– Field Picture	-64:63.5
Max. Frames/Second	30	Levels Permitted	Main
Chroma Format	4:2:2 or 4:2:0	Upper Bound Luminance Sample Rate	11,059,200 (4:2:2)
Picture Coding Type	I,P,B	Upper Bound for Bit Rate	50 Mb/s
Scalable Mode	—	VBV Minimum Size Required	9,437,184 bits
Intra-Coded DC Precision	8,9,10,11		

6

The Roots of Storage and Transmission

In our last two installments, we have gone on a foray into the thicket that surrounds advanced television, hoping to concentrate on the forest, for a while, instead of the trees. After first checking our map, we have so far stopped in the clearings in which are found users and applications, sources, source coding, and channel coding. We have also looked at the groves for the several different modulation schemes. This time out, we will visit the stands where the varieties of storage and transmission are to be found.

Storage and transmission are the real reasons why we do to the video and audio all the things we have been describing so far. When we digitize images and sounds, the amount of storage space or bandwidth required grows many times compared to what is needed for the original analog signals. When we then compress the digital representations of the signals, we can manage to squeeze them into significantly less space than the analog versions. This, in turn, leads to new ways to store and transmit signals that were not before practical and is what now opens new possibilities and opportunities for older lines of business and will lead to wholly new businesses in the not-too-distant future.

STORAGE – DIGITAL VIDEOTAPE

There are two fundamental reasons for storing compressed programs: for the direct distribution of programs to the end user using the media on which they

---

First published November, 1994. Number 24 in the series.

are stored, and for indirect distribution through other media for further processing or delivery to the end user. Direct distribution is through "packaged media" such as videotapes and videodisks. New types of equipment for digital versions of these media are currently in development and will be available in both consumer and professional forms before too long.

Digital videotape formats are being developed by groups centered in the United States and Japan involving over a dozen consumer electronics manufacturers to handle high definition (HDTV) and standard definition (SDTV) signals, respectively. There is considerable commonality but apparently, as of yet, no unification of the details. The target market is the consumer, but versions supporting professional interfaces and features are expected to be built on the same platforms and may actually be introduced first.

The formats both have a capacity to record 25 Megabits per second (Mb/s) on 6 mm tape. The same cassettes are used by both and come in small and large sizes. The small cassette stores around 13 Gigabytes and runs for a bit over an hour. The larger cassette stores about 43 Gbytes and runs for nearly 4 hours.

The HDTV version is intended for recording and playing back Grand Alliance signals. It avoids the use of an encoder or decoder, handling only the MPEG-2 Transport Stream packets of the GA system on its input and output. This should make it a relatively inexpensive machine. The incoming signals would most likely undergo forward error correction (FEC) processing and the addition of new error correction coding (ECC) tailored to the tape environment.

Aside from the change in ECC, what is recorded on the tape is likely to be somewhat different from the interface signals in order to permit "trick" modes of operation. For example, a proposal from Hitachi last year involves spreading duplicate copies of the independently-coded frames (I-frames) in special tracks at a steep angle to the normal track angle. These stunt mode tracks are interspersed with the data of the normal tracks, following the path of the head scanner when the tape is running at high speed. This permits picture-in-shuttle by allowing recovery of just the I-frames. Under this proposal, shuttle speeds would be at fixed ratios to normal forward play speed that depend on the total number of frames in the group of pictures (GOP).

The SDTV version has possibilities for several new and unique features that could differentiate it from previous VCR designs. With its 25 Mb/s data rate, it could offer the possibility to simultaneously record four or more compressed signals, so long as their data rates are 6 Mb/s or less. It could also offer the prospect of downloading material at faster than real time rates, transferring in 20 minutes a two-hour movie coded at 4 Mb/s, for example. Consideration is likewise being given in the format design to accommodating the needs of camcorders that do not have a complete MPEG encoder, permitting use of the full 25 Mb/s for a more lightly compressed signal.

STORAGE – DIGITAL VIDEODISK

Two new variations on the compact disc theme are being developed by groups comprised of Sony Corporation and Philips Electronics on one hand and Toshiba Corporation and Time Warner, Inc., on the other. These high-density compact discs have two or more times the density of standard discs. Since both the track pitch and the spacing along the track are cut in half, the amount of data recorded is essentially four times that of current CDs. The discs run at two times the speed of current compact discs. With digital compression of the video and audio, something over 2 hours can be recorded on the high density disks. The discs will likely use MPEG-2 coding or something closely related, but these details have not yet been released by either group.

The Sony/Philips approach is closely related to the current technology of compact discs. Presumably the mastering would require some new techniques to place the pits that carry the data closer together, and somewhat cleaner rooms and tighter tolerances would be needed in the replication process. But the technology and the equipment to duplicate the material would remain fundamentally the same as for compact discs. In addition, the patents covering compact discs, on which Sony and Philips hold a virtual lock, would remain in full force over the new discs along with whatever new patents the two companies have obtained for the extensions to the technology.

The Toshiba/Time Warner scheme is claimed to provide considerably more storage capacity that can be retrieved at a higher data transfer rate than is provided by the Sony/Philips system. The results of these improvements can be traded off between longer storage time and higher image quality. Better quality can be achieved at higher rates, but that eats into the increased storage space that could provide even longer capacity at lower rates. The Toshiba/Time Warner system is reported to use a two-sided disc in combination with high density recording to achieve its greater capacity. If this approach were adopted, the patent rights of the new scheme would add to or supplant some of those of Sony and Philips, thereby diluting their control over the medium.

The Hollywood studios are currently discussing which of the two approaches to use. There is a distinct possibility here for another Betamax/VHS-style format war, with a confused consumer as the ultimate outcome.[7] This would only slow down the uptake of the technology by consumers, thus thwarting the plans of the studios to open what they hope will be a more profitable distribution channel for their libraries. They would very much like to recycle their product into a medium that they hope will bring outright sales, as opposed to rentals, as the predominant mechanism for getting material into consumers' hands.

---

[7] About a year after this was written, a deal was struck between the two system proponents and a format war averted. The result will be something of a Grand Alliance of the Digital Video Disk (DVD), drawing the best from each system.

STORAGE – SERVERS

A new form of storage that has begun to appear on the market recently is the video server. A server is a device that stores material or data on hard disks or other media and can recall it nearly instantly upon request by a host device. File servers are used in computer systems to provide high capacity central storage of files that can be accessed by any number of computers connected to a network of which the server is a part. Video servers combine file server techniques with video compression to store large volumes of video program material.

Video servers can be used for a number of divergent applications that are best implemented with different server architectures. Commercial playback for television networks and stations, for example, may require a different server topology from that needed for video-on-demand applications. Servers can be built with storage capacities ranging into the hundreds of hours and with output channels numbering in the thousands. They can also be built to operate with different levels of compression and hence different levels of video performance.

Most video servers of this type are planned for use with MPEG-2 compression. Proposals so far are for use of 15 Mb/s data stream bit rates for commercial playback applications and for 1.5–3.0 Mb/s bit rates for video-on-demand applications. At 15 Mb/s, 6.75 Gbytes of storage are required for one hour's worth of commercials. At 1.5 Mb/s, this becomes 1.35 Gbytes of storage for a two-hour program or movie. Many such servers are expected to use RAID (Redundant Array of Inexpensive Disks) technology to achieve high reliability with relatively small and low cost hard drives. There are six different flavors of RAID techniques (levels 0 through 5), and they are summarized in Table 6.1.

Interfaces into the video servers shown so far have generally been in uncompressed form (raster-based signals, either NTSC-encoded or in components, either analog or digital). This has put the compression encoding process under the control of the server system designer and has allowed inclusion of switching (editing) from one program or segment to another in relatively easy fashion.

Once compression catches on in the rest of the television system and it becomes possible and necessary to input pre-compressed material, a whole range of issues that we have discussed a couple times before in this column will come into play. These issues deal with the need to standardize the encoding process for all material that is to be included in a single output stream in order to avoid buffer overflows and underflows and the "channel change" effects they imply.

On the output side, a wide variety of interfaces has been proposed. For the broadcast application, one or a few outputs in uncompressed form are currently the rule. Again, these may be NTSC-encoded or components, analog or digital. This makes program continuity integration downstream of the server relatively easy using existing switchers and other equipment. When the output must be

Table 6.1 – Redundant Arrays of Inexpensive Disks (RAID)

CHARACTERISTICS	RAID 0	RAID 1	RAID 2	RAID 3	RAID 4	RAID 5
Data on original disk duplicated or mirrored on second disk		✓				
Data striped across multiple disks using one byte per drive accessed	✓			✓		✓
Data striped across multiple disks using full sectors on each drive			✓		✓	
Error detection and correction codes stored on a separate check disk			✓	✓	✓	
Parity interleaved with data and striped across several disks						✓
ADVANTAGES	Increased speed	Full redundancy of data	Large data block efficiency	Increased speed Large data block efficiency	Increased efficiency for small data blocks	Allows multiple simultaneous writes
DISADVANTAGES	No error detection/correction	Only 50% of disk capacity usable	Unnecessarily redundant error detection/correction	High overhead working with small amounts of data	Slow writing of data due to shared check disk	Most complex controller required

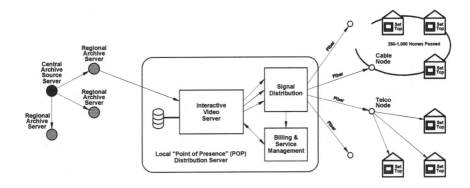

Figure 6.1 – Hierarchical Video Server Architecture

in compressed form sometime in the future, the same issues just discussed regarding the input will apply to the output as well.

For video-on-demand and near video-on-demand applications, very large numbers of simultaneous outputs are envisioned. These may be in a heirarchical architecture with a source server feeding distribution servers that store and forward the programs to viewers who request them. (See Figure 6.1.) Several different architectures have been proposed for servers having large quantities of outputs; a generic version of these is shown in Figure 6.2.

One way to achieve the numbers of outputs that will be needed is through the use of wideband channels that carry many multiplexed program data streams. If, for example, each program is encoded in 1.5 Mbits/second ($\approx$T1 rate), then 28 of them can fit in a single 45 Mb/s (DS-3, OC-1, or STS-1)output channel. Combined with ATM for routing, this can allow large numbers of outputs to be produced with manageable amounts of hardware. (We will discuss

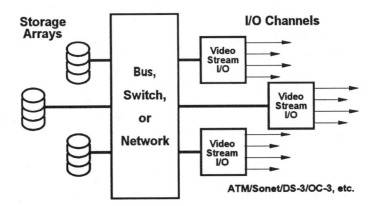

Figure 6.2 – Generic Video Server

T1, DS-3, ATM, etc. briefly later in this overview and in more detail in a future column.)

STORAGE – EDITING SYSTEMS

Storage of compressed images for editing purposes is, of course, intended to take advantage of the opportunity for non-linear editing provided by storage on hard disks and the possibility of allowing several editors to work on the same material at one time through computer networking. The use of digital video compression makes it possible to store reasonable amounts of material at an economical cost.

Until recently for the most part, such systems have been for off-line use, with on-line conforming required after editing decisions are made off-line. Recent advances have made on-line editing possible, either through use of larger, faster storage systems that do not require compression or through use of better compression methods that offer higher output quality.

While they produce excellent pictures and can be quite elegant in their designs, we will not go into detail on the non-compressed systems since they are somewhat more obvious in their technologies than the compressed systems.

As opposed to video servers, which will mostly use MPEG-2 compression, editing systems mostly use other forms of compression, for now at least. With a classic MPEG-2 implementation, a group of pictures (GOP) is on the order of 12 frames long. Splicing, as required for editing, can only occur at an entry point to a GOP, i.e. at the start of its independently-coded, or intra-frame-coded, frame (I-frame). This is not sufficiently frequent for editing, which ideally should be possible on every frame.

The choices, then, become decoding and re-encoding the material so that editing can take place in the non-compressed domain, or using some form of compression that permits splicing on every frame boundary. The first of these choices is not desirable because the image quality will be reduced as a result. The implication of the latter choice is that the image must be compressed completely with intra-frame-coded frames. In general, this is what is done.

Most of the early developments in non-linear editors based on compression used a compression scheme called JPEG. JPEG, which stands for Joint Photographic Expert Group, is a system for compressing the data representing still images and is based on use of the Discrete Cosine Transform (DCT). Each frame is treated as a still image and compressed separately – just what is needed for editing. Because the original JPEG method only provided for compressing single images, as more editing systems were developed, eventually Motion JPEG was created to handle sequences of images. Ultimately, JPEG techniques formed the basis for I-frame compression in the MPEG standards.

A newer approach to non-linear editing uses a different compression method altogether called Wavelet compression. Wavelets are a form of sub-band coding in which a succession of filters is applied to an image in both the horizontal and vertical directions. The information within each of the sub-bands is separated and extracted for storage or transmission. The most important information is concentrated in the lower frequency sub-bands and the amount of data is reduced through selection of how many sub-bands are necessary to adequately carry an image. Wavelets served as the basis for the first "near on-line-quality" editors on the market, which have been used in selected on-line applications.

Most recently, there have been a number of efforts to develop an approach to on-line systems using the MPEG techniques. Using just I-frames or GOPs that are only two frames long, these systems expect to take advantage of the more complete facilities for handling of the compressed data provided in MPEG-2's Transport Layer. In addition, an industry effort is currently under way through SMPTE to develop a Professional Profile within the MPEG standards structure that will provide for multi-generation encoding and decoding of images for post production purposes, not just editing.

RF TRANSMISSION/WIRED TRANSMISSION

The Grand Alliance and the FCC Advisory Committee earlier this year settled on 8-VSB modulation for the terrestrial transmission of digital signals by standard broadcasters, but that is not the end of the story. Competing offerings remain in the domains of both wired and wireless cable. Even in the realm of terrestrial broadcasting, there is potential competition to 8-VSB still under development. We will look at these methods along with the myriad possibilities for the wired transmission of data in our next outing, when we will complete our walk through the forest of advanced television.

7

The Varieties of Transmission

When we began this overview of the current state of Advanced Television three episodes ago, using the metaphor of a walk through the woods concentrating on the forest instead of the trees, I, your intrepid tour guide, expected that we could cover the terrain in two installments. That it will actually take us four outings to cover the ground even superficially is a strong indicator of the vastness and complexity of the realm.

As a reminder of where we have been, we first checked our map, then we stopped in the clearings in which are found users and applications, sources, source coding, and channel coding. We next looked at MPEG-2 levels and profiles and at the groves for the several different modulation schemes. Last time, we saw four different varieties of storage. Finally, this time out, we will look at two categories of transmission: wired and wireless.

WIRED TRANSMISSION

There are two fundamental applications of wired transmission of digital signals with which those of us involved in Advanced Television will be concerned in the future: transmission for delivery of material and programs and transmission for distribution of programs. Delivery, as used here, means the delivery of segments and complete programs from sources such as networks to distribution points such as network affiliates and cable headends. Distribution means the

First published January, 1995. Number 25 in the series.

dispersion of programs and interstitial continuity elements to the ultimate user or viewer.

There are numerous techniques that can be used for the wired transportation of digital signals. Since we are looking at the forest and not the trees, we will just be able to discern their basic features and see a few of their characteristics. We are now in a neck of the woods we haven't visited before, but we will save for the future a detailed look at each of them.

An important factor that differentiates some of the techniques we are about to describe is that they are suitable solely for delivery while others are suitable for both delivery and distribution. We will describe the various methods and also look at where they might be applied.

WIRED RF TRANSMISSION

We have spent a great deal of space in this column describing the various methods for carrying digital signals on radio frequency carriers. While much attention in the industry has gone to making rf transmission work for transmission through the ether, rf transmission is just as significant over both copper and fiber optic cables. The use of rf techniques over cables is primarily applicable to distribution applications as opposed to delivery systems.

Copper Pairs

Working our way from the lowest to the highest capacity cables and the narrowest to widest bandwidth techniques, we start with the lowly copper pair. Copper pairs, of course, are the ubiquitous conveyance for bringing telephone service to over 100 million locations within the United States alone. Over a period of several decades, techniques have been developed to permit copper pairs to carry more than just the single telephone circuit with which they sprouted. This resulted from the decreasing cost of electronics and the concurrent increasing cost of the copper and particularly of its installation.

Initially, methods emerged to permit multiplexing several voice channels on a single pair using analog frequency division multiplexing (FDM) methods. By locating demultiplexers near subscribers, this allowed more customers to be served without pulling more cables. Later, a new form of telephone service was conceived that can provide each subscriber simultaneously with a digital voice channel and a digital data channel, each having a data rate of 64 kilobits per second (kb/s). This required the digital signals to be carried on low frequency rf carriers with QPSK modulation. The service is called Integrated Services Digital Network (ISDN).

From ISDN techniques emerged extensions to higher rate digital techniques called HDSL (High-speed Digital Subscriber Line) and ADSL (Asymmetric Digital Subscriber Line). HDSL provides approximately 750 kb/s of data bandwidth in each direction by using more complex modulation than ISDN (essentially 16-QAM), yielding about ½-DS1 rate (about which more in the next section). ADSL takes all of the bandwidth and puts it in one direction

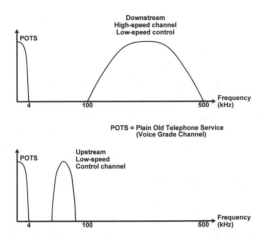

Figure 7.1 – Asymmetric Digital Subscriber Line (ADSL)
Downstream and Upstream Frequency Spectra

(downstream – toward the subscriber), with a relatively narrow upstream channel for return messages. See Figure 7.1 for a diagram of the spectrum of ADSL. The modulation used for ADSL can be either 16-QAM or a version of OFDM, depending upon the vendor. An enhanced version of ADSL is expected to carry as much as 6 Mb/s of data bandwidth.

ADSL has been seen by some local telephone companies (Local Exchange Carriers or LECs), particularly Bell Atlantic (BA), as a means to begin delivery of video services to the home without requiring the installation of new wiring from the central office (C.O. or Serving Wire Center, in telco parlance). In fact, BA is using ADSL to support a market test of video-on-demand services in northern Virginia using MPEG compression to 1.5 Mb/s. But after some initial enthusiasm for ADSL, the telcos have, in general, soured on ADSL because of the high cost of the home terminals (set-top boxes) and corresponding C.O. equipment that are required.

Coaxial Cables

Next up the ladder of cable types is coax. Coax, certainly, is the ubiquitous distribution medium for cable television operations, at least for the final distance to the subscriber. Coax started its life in cable systems as the sole conveyer of television signals, with a trunk and root system that included cascades of trunk cables and amplifiers, bridger amplifiers, feeder cables and amplifiers, taps, and drops to the home or other customer premises. This is shown in Figure 7.2.

Coax systems started with barely enough bandwidth for 12 channels but grew over time to accommodate first 36 channels (with 300 Mhz bandwidth) then 60 channels (based on 450 Mhz bandwidth) using their initial topology.

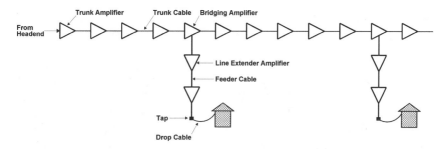

Figure 7.2 – Typical All-Coax Cable Plant Layout

Recently, in combination with fiber optic trunking (about which more in a moment), they have moved to bandwidths of 550 MHz and 750 MHz, supporting about 80 and 110 channels respectively. Ultimately, coaxial cables are seen as being capable of delivering 1 GHz bandwidths so long as they are restricted to the last couple hundred feet to the subscriber.

RF distribution of digital signals over coaxial cables can be by any of the methods that have been discussed for terrestrial broadcasting, namely QAM, VSB, and COFDM. COFDM offers no value for cable transmission since its principal claimed benefit is operation in the presence of multiple 0 dB echoes. Given its expected much higher complexity, COFDM is unlikely to see use on cable unless it is adopted for broadcasting, in which case it would presumably be used only to carry broadcast signals.

As between QAM and VSB, there are advantages and disadvantages for each. The choice between them can be made on a system-by-system basis and will probably depend primarily on each operator's selection of set-top box vendor. Among the major set-top manufacturers, General Instrument (Jerrold), Scientific Atlanta, and a number of others are developing 64-QAM, with 256-QAM a possibility for higher data rates; Zenith is developing 8-VSB, with 16-VSB as a higher bandwidth option. Since each of the proposed systems will use standard MPEG-2 source coding and transport stream packetization, either modulation method can be used successfully with the precompressed programming that can be expected to be delivered by satellite to cable headends.

An important decision for cable operators is how to allocate the spectrum on their systems between analog channels and digital channels. Since they cannot change all of their set-top boxes overnight, nor would they want to because of the costs involved, there will have to be a lengthy transition period during which both analog and digital will be in use.

A proposal currently in some favor within the cable industry is to use from 54 up to 550 MHz for downstream analog signals, from 550 to 750 MHz for downstream digital signals, and from 750 MHz to 1 GHz for both downstream and upstream digital signals for non-entertainment purposes. Also possible for upstream signals with certain system topologies is the low frequency range from 5 to 50 MHz that didn't work well in the past, but that story will have to wait for another time.

The arrangement just described would provide just over 33 channels of 6 MHz bandwidth, capable of carrying over 165 compressed programs of full motion video with "near broadcast quality" and double that number of movies with "near VHS quality." Just as important, this arrangement provides for the many expanded services cable operators hope to offer when they begin their competition with the telecommunications industry by providing telephony and other services.

Fiber Optics and Hybrid Fiber/Coax

Over the last half dozen years or so, the technology of laser modulation has improved to the point that a full complement of channels can be carried on a fiber with very reasonable distortion levels (of composite second order, CSO, and composite triple beat, CTB). Because of their small size and light weight, many fibers can be installed for nearly the same cost as a single fiber, with the installation cost dominating over the material cost. These factors have enabled a revolution in the outside plant design of cable systems. The revolution is termed "hybrid fiber/coax."

Fiber initially began to appear in cable systems in an arrangement called "fiber super trunking." In this scheme, the long cascades of coaxial cables and trunk amplifiers were replaced with very short amplifier cascades that involved splitting up the trunk cable and turning half the amplifiers around. The shortened cascades permit much improved picture performance for the analog signals through reduced noise addition and distortion. This, in turn, allows wider bandwidths to be achieved through the amplifiers that remain. The use of fiber then permits the center of each trunk segment to be fed with signals virtually as good as those at the headend. This arrangement is shown in Figure 7.3.

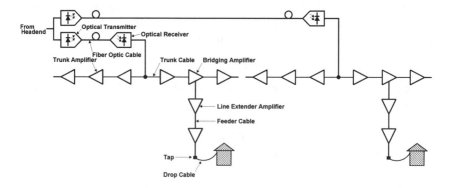

Figure 7.3 – Initial Hybrid Fiber/Coax Cable Plant Layout Based on Fiber Super-Trunking

With all the benefits that derive from using fiber to bring signals closer to the subscriber through super trunking, it was only natural for the concept to be extended. Later versions are termed "fiber to the feeder" and "fiber to the node." In these instances respectively, the trunk is replaced completely and the signals delivered directly by fiber to the feeder cables, and the feeder cables are replaced by nodes that feed numbers of subscribers that can grow progressively smaller over time as the nodes are subdivided. The nodes might start with 2000 subscribers each and after substantial subdivision end up with 250 or even 125 subscribers each.

The benefit of small nodes is that, ultimately, each subscriber could have dedicated bandwidth (assuming 5:1 compression in terms of programs, 250 subscribers at the equivalent of 1.2 MHz each need only 300 MHz) thereby permitting video-on-demand and other interactive services. Combined with the capability for upstream transmission, this would give cable operators the ability to deliver all of the services proposed by the telcos for delivery through "fiber to the home."

In fact, this approach is so attractive economically that the telcos have themselves adopted hybrid fiber/coax for their future distribution networks to consumers. The first such system is now being installed by Bell Atlantic in Toms River, NJ.

WIRED BASEBAND TRANSMISSION

For quite a few years, the rallying cry among telephone companies was fiber to the home (FTH). But fiber to the home is a technique fraught with many problems. Among these are the problems of powering the equipment that must be on the end of the fiber to extract the signals and convert them back to an electrical form to interface with the telephone, the television, and the computer. Powering this equipment during power outages for what is termed "lifeline service" is a particularly difficult challenge.

Exacerbating the matter, the cost of rewiring every home with fiber and buying the more expensive equipment required to deal with fiber makes FTH a very hard sell when no one is sure that consumers will be willing to pay much extra for the additional services. This has led to the abandoning of FTH, at least for the near term, and the adoption of hybrid fiber/coax by the telcos, as was just mentioned.

Thus wired baseband transmission is likely to see its primary application in delivery systems rather than distribution services, at least in the near-to-mid term. Nevertheless wired baseband is a very significant technique that builds on the interconnection methods of the telephone networks to carry video as well as voice and data to the many distribution points.

Hierarchical Multiplexing

The basis for virtually all of the telephone network digital interconnections in North America is the 64 kb/s "toll quality" voice channel. This service starts

Table 7.1 – Asynchronous Digital Hierarchy (ADH)

Level	Base Rate	Multiplier	Payload	Overhead	Total
DSØ	64 kb/s	1	64 kb/s	0	64 kb/s
DS1	DSØ	24	1.536 Mb/s	8 kb/s	1.544 Mb/s
DS2	DS1	4	6.176 Mb/s	136 kb/s	6.312 Mb/s
DS3	DS2	7	44.184 Mb/s	552 kb/s	44.736 Mb/s

with 8 kHz sampling of the voice and applies compression to reduce the data to 8 bits per sample. The result is called a DSØ (pronounced Dee Ess Zero) digital signalling channel.

Now that we have a voice channel, it would be nice to be able to combine that channel with other such channels in order to transmit them as a group through a circuit. This is done by time division multiplexing (TDM) the channels in a hierchical manner, as shown in Table 7.1. At each level of the hierarchy, a multiple of the channels from the next lower level is assembled and some overhead is added to synchronize the demultiplexing function. For arcane reasons related to timing and synchronization, in telco parlance, this scheme is called the "asynchronous digital hierarchy" (ADH).

As we move up the hierarchy, we combine 24 DSØs to form a DS1. Twenty-four DSØs add up to 1.536 Mb/s to which are added 8 kb/s for a total DS1 rate of 1.544 Mb/s. (This is often called the "T1" rate, something of a misnomer since the T in T1 refers to a twisted pair connection.) Four DS1s can be combined for 6.176 Mb/s, to which is added 136 kb/s of overhead to make a DS2. Seven DS2s, in turn, can be multiplexed at a rate of 44.184 Mb/s to, with the addition of 552 kb/s of overhead, yield a DS3.

It is important to note in this hierarchy that what is overhead at a lower level becomes payload at a higher level. Thus if a signal enters the multiplex at the DS2 level, the payload rate that can be used is 6.312 Mb/s – the total of the lower level signals and the DS2 overhead. Although not a perfect fit for the data rates that may be used for compressed digital video transport streams, use of data channels with standard telco data rates through "bit stuffing" and "packet stuffing" may offer advantages in the ability to use standard equipment and to do switching and multiplexing through a common carrier's network.

Optical Networks

A similar multiplexing regime exists for combining signals for optical transmission of baseband data. In this instance the lowest level is one that is appropriate for carrying a DS3 signal, and it is called OC-1 (Optical Carrier level 1). In reality, OC-1 provides a payload capacity of 50.112 Mb/s and adds overhead of 1.728 Mb/s for a total of 51.84 Mb/s. The interesting fact about the optical multiplexing scheme, however, is that at all levels higher than OC-1, there is no additional overhead required. The demultiplexing function can be carried out using the overhead from the lower levels. This is based on the fact that the optical hierarchy is a "synchronous digital hierarchy" (SDH).

The optical network heirarchy is based on simple multiples of OC-1, with normal levels in the hierarchy specified in multiples of 3 times OC-1. Thus the standard optical network rates above OC-1 are OC-3, OC-6, OC-9, OC-12, etc., as shown in Table 7.2. In addition, there are electrical interfaces at rates equivalent to the two lowest optical rates in order to permit connection into the system. These electrical rates are STS-1 for the electrical interface equivalent to OC-1 and STS-3 for the electrical interface at OC-3 level (155.52 Mb/s). At higher levels, all of the connections are made optically, with multiplexers and switches providing conversion to electrical form and back to optical form before and after their respective functions.

Above OC-12, it is normal for multiplexers to work at OC-48 rates and powers of two times OC-48 (OC-96, OC-192, etc.). Equipment at OC-48 (2488.32 Mb/s) is now the norm, OC-96 (4976.64 Mb/s) is achieveable but still expensive, and OC-192 (9953.28 Mb/s) is in the prototying stage. Clearly fiber optics offers enormous potential bandwidth.

Table 7.2 – Synchronous Digital Hierarchy (SDH)

Level[1]	Base Rate	Multiplier	Payload	Overhead	Total
OC-1	DS3+[2]	1	50.112 Mb/s	1.728 Mb/s	51.84 Mb/s
OC-3	OC-1	3	155.52 Mb/s	0	155.52 Mb/s
OC-12	OC-1	12	622.08 Mb/s	0	622.08 Mb/s
OC-48	OC-1	48	2.48832 Gb/s	0	2.48832 Gb/s
OC-96	OC-1	96	4.97664 Gb/s	0	4.97664 Gb/s
OC-192	OC-1	192	9.95328 Gb/s	0	9.95328 Gb/s

Notes: 1. Levels shown are the most commonly used and the only ones for which equipment is expected to be generally available in the near future. (Currently, up to OC-48 is common, OC-96 is in production, and OC-192 is in development.) All other multiples of OC-1 are permitted but will only become available if manufacturers find enough support for them among users to make their production financially attractive.

2. The basic OC-1 payload rate includes sufficient data bandwidth to carry a DS3 signal plus other signals that total up to 5.376 Mb/s. Any portion of the additional capacity not used for carrying data is "bit stuffed" to make up the total channel rate.

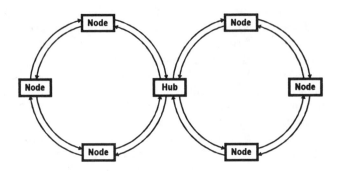

Figure 7.4 – SONET Rings Connected at a Hub

SONET

An extension of the optical hierarchy is the Synchronous Optical Network (SONET). SONET takes the basic multiplexing structure and adds some important features that make it a very powerful networking function. The basic arrangement of a SONET network is a ring structure as shown in Figure 7.4. The ring consists of two fibers with signals travelling in opposite directions on them. With this scheme, the ring can be cut at any point, and full communication can be maintained. In general, SONET networks are quoted as recovering from either a fiber cut or equipment failure within 50 milliseconds.

Other benefits of the SONET structure are efficient management of bandwidth through the ability to drop and add channels at any of the network nodes, the ability to add or delete nodes or hubs while the network remains in service, and the ability to expand capacity in multiples of four (starting at OC-12) with the network still in service. Yet another important facet of SONET is its synergy with ATM.

ATM

ATM stands for Asynchronous Transfer Mode. It is a mechanism by which signals can be transported without having to be synchronous with one another, i.e. without having to fit into one of the hierarchical data rates we've been describing. It is a system for carrying signals in packets (ATM "cells") for purposes of transmission, switching, and routing. Interestingly, ATM cells can be carried on any of the hierarchy levels just discussed.

ATM works by chopping up message traffic into 48-byte blocks to which it adds 5 bytes of header data for a total of 53 bytes (octets) per cell. The 5 byte header tells nothing of the content of the cell but rather defines its source and its destination. Through this means "virtual channels" are created. Virtual channels can be set up on a semi-fixed basis or they can be established and taken down as needed on demand.

An important factor about ATM is that it is intended to provide a means for carrying video, voice, and data on the same underlying circuits. Since each form of data gets the bandwidth it needs through the allocation of packets to it, in theory all can share a channel without being aware of the presence of the others. While this is a nice theory, some care is needed to make certain that the desired separation of services actually takes place.

Both video and audio services require a relatively constant data rate if they are to succeed without large buffers and without glitches or breakup. The video and audio signals are said to require "isochronous" (constant time) delivery. Since many of the signals with which they are intermixed in an ATM switch or channel are bursty in nature, it is possible for the switch to become "congested," that is for its peak capacity to be exceeded. When this happens, it must either delay or drop packets. It is then up to the receiver of the signal to recognize the loss and take whatever corrective steps are appropriate, such as requesting that they be resent by the sender.

This approach generally will not work for video and audio signals. Therefore it is necessary to provide a mechanism within the ATM system to identify signals that cannot withstand the dropping or delaying of its cells so that isochronous delivery can be assured. This is an area currently getting a lot of attention from the organizations developing ATM.

ATM has been proposed as a means for delivering programming all the way to the home through various cable and telco network designs. Whether delivered by rf or baseband means, it will require that adequate bandwidth be accessible from the neighborhood so that each network user can receive service unblocked by the activities of nearby users.

WIRELESS TRANSMISSION

There are three places where wireless transmission will be used: terrestrial broadcasting, satellite broadcasting, and wireless cable. Terrestrial broadcasting is working to set standards. Satellite broadcasting doesn't have many options because of the nature of satellite transponders. Because of its lack of expected standards and its many degrees of freedom, wireless cable presents the more interesting situation.

Terrestrial Broadcasting

Terrestrial broadcasting, as we all know, has been working with the FCC, through the Advisory Committee on Advanced Television Service and other organizations, to develop a single standard for delivery of advanced television signals to the home and elsewhere. Following the determination in February, 1993, that the U.S. system would digital, there was a coalesence of the four remaining system proponents into a Grand Alliance (GA) the goal of which was to use the best of each of the proposed systems to develop a single, unified approach that represented the "best of the best."

The GA chose to use the MPEG-2 standards, with as little modification or constraint as possible, to define the video compression and the transport layers of its system. It also chose the Dolby AC-3 system for the audio compression element of the system, and it selected Zenith's 8-VSB approach for the transmission subsystem after a laboratory "bake off" between that technique and 32-QAM. The 8-VSB transmission subsystem has subsequently been very successfully field tested in Charlotte, NC. We have discussed many of these matters in considerable detail on previous occasions and will not repeat the examination here.

There are two important new wrinkles with regard to terrestrial broadcasting. One is that consideration is being given by some in the broadcasting industry to the use of COFDM in place of 8-VSB for the transmission subsystem. There is a group of broadcasters who have put up some funding to have a prototype built and tested, with the expectation that, if it proves to be significantly better than 8-VSB, it will be proposed to the FCC as an alternative. It is unclear at this point how cooperative the Grand Allliance would be with regard to integrating such a subsystem with the remainder of its system.

The other new wrinkle is the effort on the part of many broadcasters to obtain the ability to use any new digital transmission capability to carry multiple standard definition television (SDTV) signals in addition to or in lieu of the HDTV signals for which the system was originally intended. Because of the use of MPEG-2 methods, the use of either type of signal or their intermixture is technologically insignificant. Much more at issue is whether broadcasters will be permitted to use their digital transmissions in the ways they want through operation of the rules the FCC adopts, any actions that may be taken by Congress, and the impact of the Supreme Court's 1949 Ashbacker decision that affects the ability of government agencies (i.e. the FCC) to limit access to newly available resources to designated applicants.

Satellite Broadcasting

One of the principal design goals for Direct Broadcast Satellite (DBS) systems is the use of very small receiving antennas. A relatively large amount of bandwidth is available (500 MHz total) at the Ku band frequencies involved. Consequently, the digital modulation scheme used is a very low order system that requires a large bandwidth for a given data rate but provides a very low threshold sensitivity. Thus the modulation method of choice for DBS, as for virtually all digital satellite transmission, is QPSK.

There is currently only one operational DBS system, the Hughes system that carries DIRECTV ™ and USSB, and so that is the only system for which we can look at some of the other characteristics. It would seem likely, however, that other DBS operations will follow the same model.

The receiving systems for DIRECTV™/USSB are currently manufactured exclusively by Thomson Consumer Electronics under the RCA label and marketed as the Digital Satellite System. Once Thomson has sold 1 million of the receivers, Sony will become a second source. Later, there will be other manufacturers.

The receivers now being sold are capable of "near MPEG-2" operation. The reason for the "near" notation is that the full MPEG-2 standard was not completed at the time the design had to be frozen. Thus the Thomson receivers can handle MPEG-2 video and audio compression but use a proprietary transport layer.

In general, MPEG-2 decoders can also decode MPEG-1 signals. DIRECTV™/USSB began operation before the software for their almost-MPEG-2 encoders was available, and so they used an enhanced version of MPEG-1 encoding to get on the air. All of the receivers deployed are capable of decoding the MPEG-2-type signals when they are put on the air and will automatically follow whatever the encoders send them.

MPEG-1 creates a number of artifacts in the compressed image that MPEG-2 avoids. Consequently, in these early days of DIRECTV™/USSB operations, some artifacts are visible. These can be expected to be significantly reduced or eliminated when the MPEG-2 encoding process is installed. The upgrade is currently planned for the first quarter of 1995.

Wireless Cable

Wireless cable uses microwave transmission at 2.5 GHz to distribute cable programming to subscribers over as many as 33 channels. Its transmissions behave much like those of terrestrial broadcasting, but it also has many of the characteristics of wired cable. In a full wireless cable system, the microwave signals are transmitted on adjacent channels using a common antenna or a pair of antennas for alternating channels to make combining them easier.

The wireless cable signals are subject to the same echoes, multipath, and signal distortions as their lower frequency counterparts. In many cases, the severity of these anomalies is much less, however, because of the frequency involved. Nevertheless, the set-top boxes used for wireless cable will require fairly capable adaptive equalizers more akin to those for terrestrial broadcasting than to the simpler ones for wired cable.

Unlike terrestrial broadcasting and more like wired cable, it appears that the FCC has no interest in setting standards for digital transmission on wireless cable. Thus wireless cable operators, like their cable cousins, will be able to select the modulation schemes they will use based upon their selections of set-top box vendors. This leads to one of the more interesting aspects of advanced television implementation for wireless cable since, unlike wired cable, wireless cable transmissions are subject to co-channel interference.

Since there can be no co-channel interference on wired cable and since terrestrial broadcasting will have a single standard, wireless cable is the only place where co-channel interference will occur between different modulation schemes. The consequence of this is that additional testing may be required to make sure the correct values are chosen for interference protection ratios between neighboring operations using different forms of modulation. It will be fascinating to see whether the interference caused to one another by signals of different modulation systems is the same as the interference between signals of the same system.

TRAIL'S END

Well, we've come to the end of our treck through the forest of advanced television. I hope we've met our goal of concentrating on the forest and ignoring the trees, thereby providing an overview of the technology and its players. Next time we'll go back to looking at the trees. Until then ... happy trails!

PART 3
Digital Video Compression

Digital Video Compression lies at the heart of Advanced Television as we use the term. Without it, there would be no Advanced Television in this sense. Thus one of the first areas we explored in any depth was video compression. It took five articles, spread over the period from October, 1992, through February, 1993, appearing here as Chapters 8 through 12, to look at the basic techniques. The intent was to make the concepts accessible without getting involved in a rigorous theoretical treatment. Had this material appeared in a text book, it might constitute a single introductory chapter on the subject.

During the period when this material was written, MPEG-2 did not exist. MPEG-1 was just becoming understood. MPEG-1 had been developed largely to permit storage of video clips on CD-ROMs and similar storage applications for very low data rates. It was nevertheless being used, often in a modified form identified as MPEG+ (or with some additional number of pluses), to encode video at considerably higher rates. One such implementation was one of the entrants in the competition for selection as the future HDTV standard by the FCC Advisory Committee, which was, after all, what Advanced Television had started out to be. How things have changed over the intervening several years!

By the time the Advisory Committee narrowed the field exclusively to digital methods, MPEG-2 was in development and was chosen to form the basis of the compression system for the new Grand Alliance. At the same time, there was recognition that the type of compression contemplated could enable many other possibilities, such as fitting multiple programs into a single channel. The race to use compression for direct broadcast satellites (DBS), cable television, distribution over copper pairs or optical fibers, packaged media, and similar methods, was also on. An increasingly wide range of uses, including most recently the digital video cassette (DVC) and the digital video disk (DVD), continue to adopt these techniques up to the present.

Notwithstanding the era when these chapters were written, the information in them remains correct today. Many of the details were refined in the MPEG-2 development, including extending its range of utility all the way down to teleconferencing and up to HDTV applications. (See Chapters 4 and 5 for more on MPEG-2 Profiles and Levels.) Yet the explanations here still cover the fundamentals of DCT-based compression systems such as the two MPEG standards. The proof of this is in the fact that I am still using essentially the same graphic materials in tutorials and seminars that I currently give on the subject. Only some of the details have been updated, as they have been in the following chapters as well.

8

Compression Means Removing Redundancy

Compression, as defined by the dictionary on the shelf, is "the act or process of shortening or condensing as if by pressing or squeezing." Compression is a word that gets thrown around a lot in discussions and articles on digital television systems. But what does it mean when applied to video? Is there a single compression technology or many? What are the techniques that compose compression? How complex are these techniques? Do the same techniques apply to compression of standard definition television and to compression of High Definition Television (HDTV)? Do the same techniques apply to broadcast television and to cable? To DBS? To wireless cable? To recorded material?

In the first installment of this series, we saw that compression is also called "Source Coding." We described source coding as including functions that sort out the information that must be sent to make a picture from the large amount of redundant information that is not required to be sent to make the picture. Source coding is usually where the largest part of the system complexity is located.

With this issue, we begin a multi-part examination of digital image compression. We will look at many aspects of compression. In each installment, we will try to start at a non-technical level and then work our way into the more technical. In this way, we can hope that the early portions of the articles will be of use to non-technical and management readers who want to have some understanding of what compression is all about. If you are a technical reader, you may wish to call the attention of your non-technical associates to this series.

First published October, 1992. Number 4 in the series.

FUNDAMENTALS OF COMPRESSION

Compression, as the term is applied to video images, takes advantage of two facts: 1. There is much in the images that is redundant (repeated, repetitive, superfluous, duplicated, exceeding what is necessary, etc.). 2. The human eye/brain combination has limitations in what it can perceive. By making use of these factors, it is possible to greatly reduce the amount of information that must be transmitted from a picture source (such as a camera) to a display (such as a television monitor or receiver) in order to convince the eye/brain combination that it is seeing an image of what the camera saw.

Why is it important to reduce the quantity of information that must be transmitted from a picture source to a display? Doing so enables many applications that are not possible otherwise, either for technical or for economic reasons. For example, compression permits the digital transmission of television images in a practical manner. An HTDV signal, in its original form, takes over 1 billion bits per second to communicate; a standard definition signal needs over 200 million bits per second. A television channel provides space for only about 20 million bits per second to be transmitted reliably. Thus the amount of data sent must be reduced on the order of 50-to-1 for HTDV. Not all of this reduction comes from compression, but the largest proportion does. With the same 50-to-1 reduction, 5 standard definition television signals can fit in the same television channel. Other reduction ratios are possible and have been proposed. Similar arguments can be made where cost of storage is the limiting factor just as available channel width is for broadcast, cable, wireless cable, and direct broadcast satellite.

The first part of the data reduction comes from eliminating some of the data to be compressed. This is done by lowering the resolution of the image in certain ways that it is hoped will not be noticed by the viewer. The eye's resolution of color is lower than its resolution of black-and-white information, for instance. Thus it is possible to reduce the amount of color information without impacting the eye's perception of a scene. This is done by filtering the color in both the horizontal and vertical directions. The effect of this is seen in Table 8.1 in the line for Net Encoder Input Video Data Rate by comparison with the line for Production Input Video Data Rate. (Non-technical readers may wish to skip the next paragraph.)

Table 8.1 shows the published characteristics of the compressed digital video systems proposed for HDTV. (Table 8.1 shows only the values for the HTDV compression systems because they are the only ones for which information has so far been published. The standard definition versions of the various proposed systems work in similar manners.) The line showing Production Input Video Data Rate is based on the use of digital video in the studio as it is practiced today, that is using a 2:1 ratio between luminance samples per line and color difference samples per line for each color difference signal, including both vertical and horizontal blanking intervals, and the like (essentially an extension of CCIR 601 techniques). The number of bits per sample or picture element (pixel) is matched to that used in the encoding process for the particular system, and the sample rate is the same as used for the encoder input. The line for Net Encoder Input Video Data Rate shows the

Table 8.1 – Data Reduction in Compressed Digital Video Systems (HDTV)

Characteristic	DigiCipher	DSC-HDTV	AD-HDTV	CC-DigiCipher
Production Input Video Data Rate (Mb/s)	858.4	1356.5 (9 bits used per pixel)	906.2	1208.3
Transmitted Scan Lines	960	720	960	720
Field Rate (Hz)	59.94	59.94 (Frame Rate)	59.94	59.94 (Frame Rate)
Interlace	2:1	1:1	2:1	1:1
Aspect Ratio	16:9	16:9	16:9	16:9
Transmitted H Freq. (kHz)	31.4685	47.203	31.4685	47.203
Active Samples/ Line (Y)	1408	1280	1500	1280
Sampling Rate (Y) MHz	53.65	75.36	56.64	75.52
Active Samples/ Line (C)	352	640	750	640
Y/C Line Ratio	2:1 (Decimation by Fields)	2:1 (Decimation by Frames)	2:1 (Decimation by Frames)	2:1 (Decimation by Frames)
Net Encoder Input Video Data Rate (Mb/s)	405.1	745.7 (9 bits used per pixel)	517.9	662.9
Compressed Video Data Rate (w/o FEC) (Mb/s)	18.24	8.46-16.92 Variable Rate	17.73	18.88
Compression Ratio	22.21:1	44.08- 88.15:1	29.21:1	35.11:1
Overall Video Data Reduction Ratio	47.06:1	80.17- 160.34:1	51.11:1	64.00:1

reduced data rate after elimination of the blanking intervals and reduction of the resolution of the color difference channels both horizontally and vertically.

Compression itself is an agglomeration of techniques, based on one or the other, or both, of the two factors mentioned previously, to prepare images for storage or for transmission to the point of use, using the minimum storage space or the minimum bandwidth possible, after which the images are decompressed for display and viewing. Different systems proposed by various proponents all use essentially the same techniques, although they use different combinations and arrangements of those techniques. They also use different parameters associated with the techniques to obtain optimal performance for their systems overall and for specific applications.

The compression schemes developed by various organizations use particular approaches to the problem of compressing video. In general, the same compression schemes are proposed to be applied to all of the applications for which each of those organizations develops solutions. Thus the same basic combination and arrangement of techniques from a proponent are proposed to be applied to compression of standard definition and of high definition television signals. The same basic compression schemes are also proposed to be applied by their proponents to broadcast television, cable television, wireless cable, direct broadcast satellite, and recorded media. The differences in the applications for the various media center on the channel coding, as was discussed in the last two installments of this column and will be again in the future.

THE NEED FOR COMMONALITY

The fact that the same compression approach can be applied across all these media has important positive implications for *interoperability* – the ability of the various media to interchange material with one another. It means that, if the same compression method is used by the several media, they can send compressed material from one to another without having to decompress (decode) it and re-compress (encode) it again. This will allow a very significant economic savings when the cost of the decoders and encoders (in particular) is considered. It can mean that the number of encoders in local television stations and in cable and wireless cable headends can be reduced meaningfully. Given the high cost expected for encoders, especially in the start-up years of ATV, this will be an important consideration.

The need for interoperability brings with it another very important consideration. If interoperability between media is to be achieved, the media will all have to adopt the same fundamental compression scheme. This results from the fact that the various compression methodologies are incompatible with one another. Thus a selection will have to be made among the several systems currently being proposed for Advanced Television. A mechanism exists for making such a selection for HDTV applications – it is the FCC Advisory Committee process, the output of which, it is assumed by many, will be used by all media.

A similar mechanism is needed for standard definition applications, and while none exists formally, steps are being taken that move in the direction of an *ad hoc*, almost-universal selection. Cable Television Laboratories (CableLabs), together with several cable operators and the Public Broadcasting Service, are currently in the

Figure 8.1 – Imaginary Television Image of a Painting

process of evaluating and selecting among half a dozen proposals for standard definition compression systems. It is anticipated that several other media will adopt whatever selection comes out of the CableLabs process. In the meantime, other user organizations are proceeding with selections of their own. If their choices are divergent from the CableLabs selection, they will be nearly locked into doing their own compression in the future, and exchanges of programming involving them will have to occur at baseband rather in compressed form. This is not the most economical approach. We will examine these economic realities in a future column.

REMOVING REDUNDANCY

Moving on to the technology of compression, there is a series of techniques that is used to radically reduce the redundant nature of moving images and to eliminate information from the picture that the eye/brain combination does not require in order to perceive the image originally captured or created. In our exploration of compression, we will start with the methods that reduce redundancy. Along the way, we will touch on some of the places that advantage is taken of the limitations of human vision. Finally, we will deal with some of the specific methods for fooling the eye/brain.

Television images generally are full of redundancy. This can be seen by thinking of a television image of a painting. (See Figure 8.1.) Once the first frame of the image of the painting has been captured, all the remaining frames that are captured and sent are redundant. What is necessary is to send an instruction that says repeatedly, "Repeat the last frame." This is an extreme example, of course, because there is no motion in the scene, but it serves to prove the point. It is also a real situation and serves as a limiting case for what is required of certain aspects of compression.

Similarly, there are other forms of redundancy. The example of the painting related to redundancy from frame-to-frame because there was no motion in the scene. Redundancy also applies to repeated elements within a single frame. In the painting of Figure 8.1, there are two identical windows. If a means can be found to send information to the display about the window on the left and then to send an instruction that says, "Make another copy of the window here" for the window on the right, it will not be necessary to repeat all of the information that would otherwise be required to describe the second window. Although this example is highly over-simplified, the concept of removing redundancy from within individual frames is appropriate to many of the techniques we will explore.

MOTION ESTIMATION/MOTION COMPENSATION

The first method for redundancy reduction used by many compression systems is predictive motion estimation combined with motion compensation. In predictive motion estimation at its most basic level, the preceding frame is stored in frame stores in both the encoder and the decoder. A prediction of the current frame based on the preceding frame is developed in identical predictors in both units. In the encoder, the prediction of the current frame is compared with (subtracted from) the

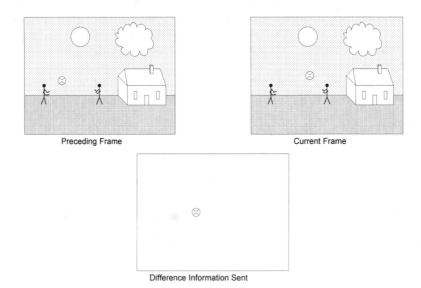

Preceding Frame   Current Frame

Difference Information Sent

Figure 8.2 – Frame Difference Information

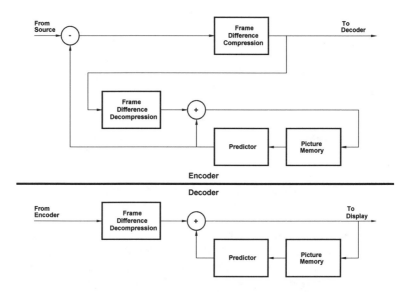

Figure 8.3 – Generic Predictive Video Compression Codec

actual current frame. Any difference between the predicted and actual frames is sent to the decoder (after further processing to remove other redundancies) to correct the decoder's prediction. This can be seen in simplified form in Figure 8.2, where the data sent represents the ball in its new position and the information needed to fill in where the ball was.

At the decoder, the correction data is added to the prediction of the current frame developed by the decoder's predictor. This creates the frame both actually to be displayed and to be stored for prediction of the next frame. At the same time, the frame difference data that was sent to the decoder is used by the encoder, through a duplicate of the prediction process in the decoder, to construct the frame to be stored for comparison with the next frame. In this way, the encoder models the decoder and sends difference data that counteracts any errors in the decoder's motion prediction. This is shown in the block diagrams of Figure 8.3.

The next level of complexity adds motion compensation to the prediction of the next frame. The image is divided into sections (blocks – about which more next chapter) that are used by a motion estimator in the encoder to try to find matching or nearly matching areas in the preceding frame. When an area in the preceding frame is found that nearly matches a section of the current frame, a motion vector is generated by the motion estimator and sent to the decoder to steer its prediction, thus making its prediction of motion much better. A motion vector indicates what part of the preceding frame should be used to predict a specific area of the current frame. (See Figure 8.4.) With better motion prediction, the amount of data that must be sent to represent the frame differences can be much less. The same motion vectors are

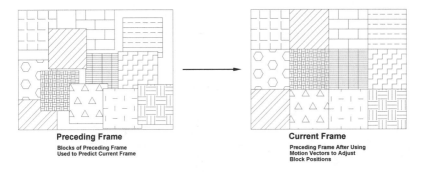

Preceding Frame

Blocks of Preceding Frame
Used to Predict Current Frame

Current Frame

Preceding Frame After Using
Motion Vectors to Adjust
Block Positions

Figure 8.4 – Use of Motion Vectors to Improve Prediction of Current Frame
from Preceding Frame

used in both the decoder, in its motion predictor, and in the encoder, in the motion
predictor that models the one in the decoder. (See the block diagrams of Figure 8.5.)

The motion vectors that are generated as part of the motion estimation/motion
compensation process must be sent along with all the other data from the encoder to
the decoder. This means that they take up data space in the channel, in what is
called a side-channel. Therefore, for them to be useful, the motion vectors must
utilize less space for transmission than would be used by the difference data they

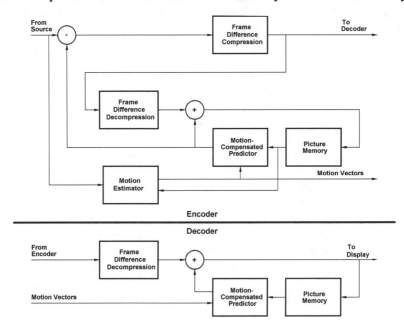

Figure 8.5 – Generic Motion-Compensated Predictive Video Compression
Codec

save. In other words, there must be a net gain in coding efficiency or compression ratio from their use. There is a fine balance that system designers must achieve in order to make their systems successful – they must optimize the data space allocated to motion vectors so as to minimize the overall data space occupied by the combination of the motion vectors and the frame difference data.

Motion compensation provides one additional feature for digital systems. So long as the standardized definitions of the motion vectors provide sufficient data space and continue to be met, improvements can be made in the future in the methods for determining the motion vectors to be sent. This will permit compatible upgrading of the quality of images delivered to the decoder without any need for upgrading of the decoder itself. This is the first of a number of forms of *extensibility* – the ability to extend the performance or features of the system without replacing decoders – that we will find as we explore digital compression systems.

Next chapter, we will continue our examination of compression with more on motion estimation/motion compensation. After that, we will begin looking at the Discrete Cosine Transform (DCT) – another almost universally applied compression technique.

9

Motion Estimation, Motion Compensation, Motion Vectors

Last chapter, we began a multi-part examination of digital image compression. We will continue in this installment, looking first at some of the non-technical implications of the technology that may be interesting to non-technical readers and then focusing on the technology itself. As before, if you are a technical reader, you may wish to call the attention of your non-technical associates to this series.

As we described it previously, digital image compression is an agglomeration of techniques that permits removal of the very large amount of redundancy that exists in all television images, with the result that only non-redundant information content of the image is sent from the picture source to the display. The principal objective of the compression process is to sort out that non-redundant content and deliver it to the display as efficiently as possible, discarding all of the redundant data generated when the image is captured or created. At the same time, there are other important, subsidiary objectives in the development of digital compression systems.

Some of these other objectives have taken on a currency in the discussion surrounding Advanced Television (ATV) that makes it important for us to examine their meaning and try to put them into context. In particular, the concepts *interoperability, scalability,* and *extensibility* have become significant causes for some parties to the debate over what the ATV standards should be. They have become industry buzz words when considering the convergence of the technologies

First published November, 1992. Number 5 in the series.

of computers and video. Since the digital technology of computers is the enabling technology for digital image compression, it is important that these concepts be given attention.

INTEROPERABILITY, SCALABILITY, EXTENSIBILITY

There are a number of sources for definitions of the ideas considered here. One important study that defined and considered the importance of these concepts was the SMPTE Task Force on Digital Image Architecture. Its final report includes definitions on which we can base our discussion. After defining the terms, we can examine their potential impacts on the television of the future.

Interoperability

Interoperability is defined as the sharing of images and equipment across application and industry boundaries. This is seen as encompassing not only the entertainment industries but all of the industries that use or could use electronic images such as telecommunications, education, engineering and science, healthcare, military and aerospace, computers and information processing. Some of the non-entertainment applications include videoconferencing, picture telephones, remote presence, distance learning, medical imaging, computer graphics, and image data bases.

To permit the sharing of images and equipment across all of these industries and their respective applications, it is necessary to have standards that they all can use. In particular, standards must exist at the boundaries of the various uses – boundaries defined as interfaces between pieces of equipment or between applications. Thus it is necessary to have common interface standards.

In the past, the entertainment industries have had their own standards, and the other industries have had their respective standards, with any commonality occuring by expedience or happenstance rather than by design. Interoperability requires the design of standards specifically for commonality, so that the images and equipment can be used in the various applications with a minimum of degradation of the images when they are passed from one application or industry to another. This means not only the use of common interfaces but also the use of standards on either side of those interfaces that are designed for easy interchange across the interfaces.

All of this, in turn, means that the future standards must be developed in a coordinated way, with an overarching plan as to how all the elements fit together. Yet two very important elements of future systems are currently being developed without the benefit of such an integrated scheme. They are the HDTV transmission system, selection of which is approaching through the efforts of the FCC Advisory Committee on Advanced Television Service, and the compressed 525-line system, selection of which is nearing completion through work led by the Cable Television Laboratories. Consideration is being given to interoperability in each of these activities, but there are no guarantees that the selections made will fit within any particular overall architecture that may be adopted in the future. This has led some to suggest, even demand, that the selection processes be stopped until an overall architecture is decided.

In the meantime, the Society of Motion Picture and Television Engineers (SMPTE) has begun an effort to define and document a digital image architecture using the recently completed Task Force report as a point of departure. It will be a several year activity, requiring substantial development of technology, including tests and demonstrations, to reach a conclusion. Its work will be all the more demanding because of the HDTV and Compressed 525 decisions that will predate it.[8]

Scalability

Scalability relates to the ability of an imaging system to adjust the level of performance by varying the amount of data that is stored, transmitted, received, or displayed. In all cases, the highest performance that can be achieved is set by the performance with which the image was originally captured or created.

Performance can be adjusted in several dimensions. The most common dimension is the resolution of the image, both in luminance and in chrominance. Other dimensions that can be varied include the fidelity of the image representation (e.g., the number of bits used for each pixel) and the image update rate (the number of frames stored, sent, or displayed per second). By permitting the varying of performance in these dimensions, scalability can help with interoperability by allowing images used in one application to be transferred to another application having a different characteristic performance level.

Examples of the use of scalability are displaying an HDTV image on a small screen, where the resolution of the image might be reduced to a level appropriate for the small screen prior to transmission or prior to display, or storing an image in very high quality form and permitting it to be recalled at any of several lower levels of resolution/performance as appropriate for the particular application. A real world instance of the latter example is the new Kodak Photo CD®, where a very high quality image is stored to permit later printing in a photographic blow-up and where lower performance versions of the same image can be extracted for presentation on a standard television screen, an HDTV display, or any other lower level application.

Extensibility

Extensibility is a feature of a system design that allows the system to evolve with advances in the underlying technologies so that additional levels of performance can be implemented without rendering obsolete those existing products that conform to the basic requirements of the system. Extensibility is thus intended to make systems "future-proof" to the extent possible. It means that the system must be structured so that devices built for the system can recognize information that they are able to interpret and use and can also ignore information that they are incapable of utilizing.

---

[8] In the end, the SMPTE effort to define a digital image architecture failed. The many disparate interests could not compromise their already hardened positions sufficiently to reach a common point of agreement.

Extensibility allows a mixture of material to be stored or transmitted together where some of the material is of a type defined originally with the system and some of the material is of a type defined later. Devices that were built before the later definitions will accept and process all the types of material they have been built to handle. Their operations will not be impaired by the presence of the later types of material, but neither will they process it.

It would thus be possible to build a system for carrying multiple compressed television images that could be displayed on a receiver designed to handle them. Later, signals could be added to the system to carry one or several daily newspapers. A receiver built for the original system would still display a selection chosen among the multiple compressed television images, but it might do nothing with the newspaper images. A receiver built after the addition was made could, in addition to receiving the television images, have the ability to present the newspapers on the screen and possibly include a printer that would allow printing selected pages, articles, or photographs.

Use of the concepts of interoperability, scalability, and extensibility in an integrated image architecture for the future offers potential benefits for all of the industries converging with one another through their use of digital imaging technology. It remains to be seen whether the ATV systems being developed under the aegis of the FCC Advisory Committee and CableLabs can fit within such an architecture and take advantage of these concepts.[9]

MOTION ESTIMATION/MOTION COMPENSATION, CONTINUED

When we left the subjects of motion estimation/motion compensation last chapter, we had introduced the idea of motion vectors sent through a side channel to help steer the prediction of the decoder, thereby improving the accuracy of the decoder's motion prediction and requiring less frame difference data to be sent. This time, we will look at how the motion vectors are developed and a more sophisticated way of arranging the frame difference information. Before we get into these subjects, however, we need to take a look at how the image is organized in a digital television system.

Each frame of a digital television image is broken down into a large number of small image samples called picture elements – pixels or pels, for short. (See Figure 9.1.) The pixels are arranged in rows and columns with the number of pixels horizontally and vertically determining the resolution of the image in each dimension. It is the number of pixels in each direction combined with the frame rate that determines the basic television system. Some typical pixel counts are given in Table 9.1.

---

[9] The Grand Alliance system and the various SDTV systems based upon MPEG-2 provide some measure of interoperability along with scalability and extensibility.

Table 9.1 – Pixel Counts for Assorted
Digital Image Structures

Pixels – H X V	System Description
768 x 483	525/59.94/2:1, 4:3
720 x 484	525/59.94/2:1, 4:3
960 x 484	525/59.94/2:1, 16:9
720 x 575	625/50/2:1, 4:3
1920 x 1035	1125/60/2:1, 16:9
1920 x 968	1050/59.94/2:1, 16:9
1280 x 720	787.5/59.94,1:1, 16:9
640 x 480	VGA
800 x 600	S-VGA
1024 x 768	8514, XGA

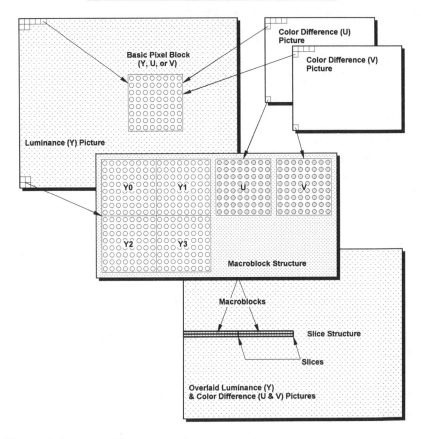

Figure 9.1 – Image Structure in Compression Systems

In order to make the processing of the images manageable, small regions of pixels are grouped together into structures called blocks. Then groups of blocks are formed for further processing. Most systems have three such hierarchical levels of picture regions for carrying out their various processing functions. Different names are applied to the regions by the system proponents, but the concepts, and in many cases the regions themselves, are similar for all of the compression systems. The names and descriptions of the regions are given in Table 9.2 for the HDTV systems. This information has not yet been published for the 525-line compression systems but will be similar. The names used in Figure 9.1 are those of the AD-HDTV system [which match those used in MPEG-2 and the Grand Alliance system and which are used throughout later parts of this book].

It should be noted that the intermediate level is generally constructed so that several blocks of luminance pixels are combined with a single block of pixels of each of the chrominance signals that represent the same area of the picture as the luminance blocks. This works because the chrominance signals have been filtered and sampled so that they have less resolution than the luminance, usually by a factor of two vertically and two to four horizontally. This is evident in Table 9.2. The only exception to this in the table is the DSC-HDTV system, which uses a somewhat more complex structure.

Table 9.2 – Structural Descriptions of Image Compression Systems (HDTV)

	DigiCipher	DSC-HDTV	AD-HDTV	CC-DigiCipher
Smallest group: Name	Block	Block	Block	Block
Smallest group: Description	8 pixels horiz. X 8 pixels vert.	8 pixels horiz. X 8 pixels vert.	8 pixels horiz. X 8 pixels vert.	8 pixels horiz. X 8 pixels vert.
Intermediate group: Name	Superblock	Superblock	Macroblock	Superblock
Intermediate group: Description	4 luminance Blocks horiz. X 2 luminance Blocks vert. +1 chrominance Block for each of U & V from same image area	2 luminance Blocks horiz. X 2 luminance Blocks vert. – Separate U & V chrominance Superblocks cover space of 2 H X 3 V luma Blocks	2 luminance Blocks horiz. X 2 luminance Blocks vert. +1 chrominance Block for each of U & V from same image area	2 luminance Blocks horiz. X 2 luminance Blocks vert. +1 chrominance Block for each of U & V from same image area
Largest group: Name	Macroblock	Slice	Slice	Macroblock
Largest group: Description	11 Superblocks horizontally	4 horiz. X 3 vert. Luminance Superblocks + 4 horiz X 2 vert. Chroma (U + V) Superblocks	13 Macroblocks horizontally	20 Superblocks horizontally

MOTION VECTORS

Now that we have established the structure for image processing, we can look at how the motion vectors are developed. Thinking back to last chapter's Figure 8.4, we can imagine how regions in the preceding frame can be identified that match regions in the current frame that are displaced from their previous locations. By sending the magnitude and direction of the displacement determined by the motion estimator in the encoder through a side channel to the predictor in the decoder, the predictor in the decoder can do a much better job, and the amount of frame difference information that must be sent can be significantly reduced.

One of the principal characteristics of the systems is the size of the regions used in finding and sending the motion vector data. This can be as small as a block or as large as the entire picture. Both extremes have been used; the Narrow-MUSE system uses one motion vector for the entire picture, while the DSC-HDTV system sometimes sends motion vectors for single blocks. It is important to note that the motion vectors sent do not have to work in a constant manner. Within the constraints of what the data structure and the decoder can handle, the methods for generating the motion vectors can change over time; they can even be adaptive to the picture content at any given instant, as implied above for the DSC-HDTV system. This all depends on the flexibility built into the initial system design.

There are a number of methods currently available for developing the motion vectors. These have names like full search and hierarchical search that describe the ways in which they identify the areas in the preceding frame that most closely match given areas in the current frame. The search methods have limits on the extent of

Table 9.3 – Motion Vector Characteristics of Compression Systems (HDTV)

Characteristic	DigiCipher	DSC-HDTV	AD-HDTV	CC-DigiCipher
Search method	Full search block matching on luminance only	Hierarchical staged search of decimated frames followed by fine search of full data	Exhaustive search matching on luminance only for P & B frames	Exhaustive search matching on luminance only
Area matched	Superblock	Block	Macroblock	Adaptive: Block or Superblock
Search range	+31/-32 pixels H +7/-8 pixels V	+47/-48 pixels H +39/-40 pixels V	+31/-32 pixels H +31/-32 pixels V	+15/-16 pixels H +7/-8 pixels V
Matching precision	1 pixel	½ pixel	½ pixel	½ pixel
Vector resolution	10 bits	Variable through complex selection process	Varies - mode data included for B frames to indicate direction of comparison	Varies with selection of block or superblock matching
Image tracking speed	0.68 frame width, 0.25 frame hght, per second	1.5 frame width, 2.7 frame height, per second	0.26 frame width, 0.33 frame hght, per second	0.75 frame wdth, 0.67 frame hght, per second

their searching and the accuracy with which they determine positional offsets. Limits on the range of search are normally expressed in terms of the number of pixels in the positive and negative directions, in the horizontal and vertical dimensions of the preceding frame, that can be searched relative to the location for which a motion vector is sought in the current frame. The accuracy with which the positional offset is determined is usually expressed as being of pixel accuracy or sub-pixel accuracy. All of this is described in more detail in Table 9.3 for the HDTV systems, which must represent all digital image compression systems since data on the 525-line compression systems has not yet been published.

MPEG

So far, we have been describing the motion estimation/motion compensation process as always proceeding from one frame to the next in sequential order. There is a major variation on that theme embodied in a digital image compression scheme developed by the Motion Picture Experts Group (MPEG) of the International Standards Organization (ISO). In fact, MPEG (pronounced "em-peg") has finished work on one system designed for relatively low data rates and is currently nearing completion of a second system for higher data rates. MPEG-style processing serves as the basis for one HDTV and a number of 525-line compression systems, several of which are listed in Table 9.4.

The important thing about the MPEG systems from the point of view of removal of temporal (frame-to-frame) redundancy is that they provide for treating the frames out of order. This is shown in Figure 9.2. In MPEG systems, there are three different types of frames that are sent from the encoder to the decoder. They are I-frames that are *intra-frame* or *independently* coded, P-frames that are *predicted* from the I-frames, and B-frames that are *bi-directionally* predicted from either past or future frames. Together, a combination of an I-frame, one or more P-frames, and some B-frames make up a Group of Pictures (GOP) that forms the basic pattern in which MPEG images are processed. There is a great deal of flexibility in MPEG in the manner in which the frames can be arranged.

Table 9.4 – MPEG-Based Systems

HDTV
AD-HDTV (ATRC)
525-Line Compression
Thomson Consumer Electronics
Philips Broadband Networks
Oak Communications
Scientific Atlanta

By permitting most of the frames in an MPEG compressed image to be B-frames, MPEG is claimed to achieve better prediction of areas uncovered by objects moving in the picture, to help uncouple prediction from coding, to average out some of the noise in the picture, and to provide a very high level of compression. At the same time, the intra-frame coded I-frames are supposed to help with random access and such tricks as slow and fast forward motion, reverse motion, still frames, and the like. Although we haven't yet explored it,

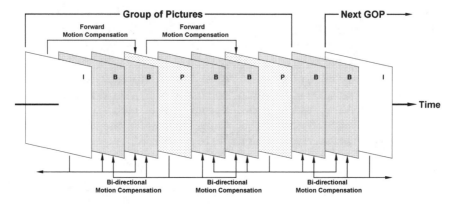

I = Intra-frame coding only, Independent frame
P = Predicted frame, forward direction, from I or previous P
B = Bi-directionally predicted frame, from surrounding I or P frames

Figure 9.2 – MPEG-2 Group of Pictures

there is also a need in all the systems to have a method for starting up the decoding process, since sending differences requires some initial reference for the differences. The I-frames provide this for MPEG systems.

In the next chapter, we will launch into an examination of the Discrete Cosine Transform (DCT), the basis in virtually all of the proposed systems for compressing the difference information that remains after the motion estimation/motion compensation process.

10

Discrete Cosine Transform

Continuing in this installment with our multi-part examination of digital image compression, we will look first at some of the non-technical implications of the technology that may be interesting to non-technical readers and then focus on the technology itself. As before, if you are a technical reader, you may wish to call the attention of your non-technical associates to this series.

Digital image compression, as we described it previously, comprises a group of disparate image and data processing techniques that enable elimination of the very large amount of redundant data that exist in all television images. The result is that only non-redundant information content of the image is sent from the picture source to the display. The principal objective of the compression process is to sort out that non-redundant content and deliver it to the display as efficiently as possible, discarding all of the redundant data generated when the image is captured or created. At the same time, there are other important, subsidiary objectives in the development of digital compression systems.

Some of these other objectives have taken on a currency in the discussion surrounding Advanced Television (ATV) that makes it important for us to examine their meaning and to try to put them into context. In particular, the concepts *interoperability*, *scalability*, and *extensibility* have become significant causes for some parties to the debate over what the ATV standards should be. They have become industry buzz words when considering the convergence of the technologies of computers and video. Since the digital technology of computers is the enabling technology for digital image compression, it is important that these concepts be given attention.

First published December, 1992. Number 6 in the series.

INTEROPERABILITY, SCALABILITY, EXTENSIBILITY

There are a number of sources for definitions of the ideas considered here. One important study that defined and considered the importance of these concepts was the SMPTE Task Force on Digital Image Architecture. Its final report includes definitions on which we can base our discussion. After defining the terms, we can examine their potential impacts on the television of the future.

Interoperability

Interoperability is defined as the sharing of images and equipment across application and industry boundaries. This is seen as encompassing not only the entertainment industries but all of the industries that use or could use electronic images such as telecommunications, education, engineering and science, healthcare, military and aerospace, computers and information processing. Some of the non-entertainment applications include videoconferencing, picture telephones, remote presence, distance learning, medical imaging, computer graphics, and image data bases.

To permit the sharing of images and equipment across all of these industries and their respective applications, it is necessary to have standards that they all can use. In particular, standards must exist at the boundaries of the various uses – boundaries defined as interfaces between pieces of equipment or between applications. Thus it is necessary to have common interface standards.

In the past, the entertainment industries have had their own standards, and the other industries have had their respective standards, with any commonality occuring by expedience or happenstance rather than by design. Interoperability requires the design of standards specifically for commonality, so that the images and equipment can be used in the various applications with a minimum of degradation of the images when they are passed from one application or industry to another. This means not only the use of common interfaces but also the use of standards on either side of those interfaces that are designed for easy interchange across the interfaces.

All of this, in turn, means that the future standards must be developed in a coordinated way, with an overarching plan as to how all the elements fit together. Yet two very important elements of future systems are currently being developed without the benefit of such an integrated scheme. They are the HDTV transmission system, selection of which is approaching through the efforts of the FCC Advisory Committee on Advanced Television Service, and the compressed 525-line system, selection of which is nearing completion through work led by the Cable Television Laboratories. Consideration is being given to interoperability in each of these activities, but there are no guarantees that the selections made will fit within any particular overall architecture that may be adopted in the future. This has led some to suggest, even demand, that the selection processes be stopped until an overall architecture is decided.

In the meantime, the Society of Motion Picture and Television Engineers (SMPTE) has begun an effort to define and document a digital image architecture using the recently completed Task Force report as a point of departure. It will be a several-year activity, requiring substantial development of technology, including tests and demonstrations, to reach a conclusion. Its work will be all the more demanding because of the HDTV and Compressed 525 decisions that will predate it.[10]

Scalability

Scalability relates to the ability of an imaging system to adjust the level of performance by varying the amount of data that is stored, transmitted, received, or displayed. In all cases, the highest performance that can be achieved is set by the performance with which the image was originally captured or created.

Performance can be adjusted in several dimensions. The most common dimension is the resolution of the image, both in luminance and in chrominance. Other dimensions that can be varied include the fidelity of the image representation (e.g., the number of bits used for each pixel) and the image update rate (the number of frames stored, sent, or displayed per second). By permitting the varying of performance in these dimensions, scalability can help with interoperability by allowing images used in one application to be transferred to another application having a different characteristic performance level.

Examples of the use of scalability are displaying an HDTV image on a small screen, where the resolution of the image might be reduced to a level appropriate for the small screen prior to transmission or prior to display, or storing an image in very high quality form and permitting it to be recalled at any of several lower levels of resolution/performance as appropriate for the particular application. A real world instance of the latter example is the new Kodak Photo CD®, where a very high quality image is stored to permit later printing in a photographic blow-up and where lower performance versions of the same image can be extracted for presentation on a standard television screen, an HDTV display, or any other lower level application.

Extensibility

Extensibility is a feature of a system design that allows the system to evolve with advances in the underlying technologies so that additional levels of performance can be implemented without rendering obsolete those existing products that conform to the basic requirements of the system. Extensibility is thus intended to make systems "future-proof" to the extent possible. It means that the system must be structured so that devices built for the system can recognize information that they are able to interpret and use and can also ignore information that they are incapable of utilizing.

Extensibility allows a mixture of material to be stored or transmitted together where some of the material is of a type defined originally with the system and some of the material is of a type defined later. Devices that were built before the later definitions were adopted will accept and process all the types of material they were

---

[10] In the end, the SMPTE effort to define a digital image architecture failed. The many disparate interests could not compromise their already hardened positions sufficiently to reach a common point of agreement.

built to handle. Their operations will not be impaired by the presence of the later types of material, but neither will they process it.

It would thus be possible to build a system for carrying multiple compressed television images that could be displayed on a receiver designed to handle them. Later, signals could be added to the system to carry one or several daily newspapers. A receiver built for the original system would still display a selection chosen among the multiple compressed television images, but it might do nothing with the newspaper images. A receiver built after the addition was made could, in addition to receiving the television images, have the ability to present the newspapers on the screen and possibly include a printer that would allow printing selected pages, articles, or photographs.

Use of the concepts of interoperability, scalability, and extensibility in an integrated image architecture for the future offers potential benefits for all of the industries converging with one another through their use of digital imaging technology. It remains to be seen whether the ATV systems being developed under the aegis of the FCC Advisory Committee and CableLabs can fit within such an architecture and take advantage of these concepts.

IMAGE COMPRESSION, Continued

Turning now to the technical side of our discussion, we dealt in the last two installments with the techniques of motion estimation and motion compensation. You will remember that these methods are used to try to eliminate redundancy in the image from one frame to the next. In Chapter 8, we used the over-simplified example of an image in which a ball was the only moving object. We assumed that the background had been sent in an earlier frame, and we were dealing with two frames that represented the ball in two different locations. To transmit the current image after sending the preceding image, we had only to send the image of the ball in its new location and the image of what had been behind the ball in its old location. These are called the frame differences.

In the last chapter, we were more realistic and recognized that in most television images there will be many objects in motion. We looked at the techniques that identify the motion and send motion vectors from the encoder to the decoder along with any frame differences or residual that result after what are called the motion estimation and motion compensation operations. The motion vectors help steer the decoder's motion estimator in the process of reconstructing the current frame. The purpose of the motion estimation and the motion compensation is to eliminate frame-to-frame (temporal) redundancy when there is motion and to reduce the amount of residual or frame difference information that must be transmitted to the receiver.

Even after the temporal redundancy is removed, there is still an enormous amount of residual data left that must be sent from the encoder to the decoder. By eliminating any redundancies that this data contains, it is possible to further compress the data and reduce the amount that must be transmitted. Several different stages of processing and compression are applied to the residual data. We will begin to look at the first of them this chapter. It will take us several installments to get through all of the major ones. Just about the time we are done with this examination, we should know which system has been selected for ATV transmission, and we can put all the pieces together in an analysis of that particular system.

TRANSFORMS

Once the motion compensation process is complete, the residual to be transmitted for each frame consists of an image matrix of pixels that exactly matches the matrix of pixels of the original image, except that it now contains only the differences between the preceding frame and the current frame that were not removed in the motion compensation process. This frame difference image is structured in just the same way as we described last time, i.e. it consists of blocks, macroblocks, and slices. The macroblock slice structures are used primarily in the motion compensation operations and in deciding some of the factors to be applied later in the data compression process. The first step in reducing the data further, however, involves just the block level.

The objective of the first step of processing of the residual is to remove any redundancy that exists at the block level. The blocks, as you will remember, consist of 8x8 arrays of pixels. What is needed is a process that can be applied to an 8x8 array, grouping together similar information into a smaller number of data points or elements. One way to do this is through the application of a mathematical formula called a transform. A transform actually yields the same number of elements after its application as existed before it was applied, but with the information organized in a different way. Thus an 8x8 array will still be an 8x8 array after transformation. But if the right transform is selected, some of the data points will carry the bulk of the information while the others will be essentially zero. An important characteristic of a transform is that it can be reversed, yielding exactly the original 8x8 array.

DISCRETE COSINE TRANSFORM

The transform that is applied to the blocks of frame difference residual in the encoder is the Discrete Cosine Transform, or DCT. A typical formula for a DCT is shown in Figure 10.1 just for interest's sake. (This one happens to be the JPEG formula.) The exact workings of the formula are unimportant for our discussion, but the results of application of the DCT form the basis for the data compression used by

$$F(u,v) = \frac{C(u)C(v)}{4} \sum_{j=0}^{7} \sum_{k=0}^{7} f(j,k) \cos\left[\frac{(2j+1)u\pi}{16}\right] \cos\left[\frac{(2k+1)v\pi}{16}\right]$$

$$C(w) = \begin{cases} \dfrac{1}{\sqrt{2}} & \text{for } w = 0 \\ 1 & \text{otherwise} \end{cases}$$

Figure 10.1 – Formula for Discrete Cosine Transform
(JPEG used as example)

139	144	149	153	155	155	155	155
144	151	153	156	159	156	156	156
150	155	160	163	158	156	156	156
159	161	162	160	160	159	159	159
159	160	161	162	162	155	155	155
161	161	161	161	160	157	157	157
162	162	161	163	162	157	157	157
162	162	161	161	163	158	158	158

Figure 10.2 – 8x8 Block of Frame Difference
Residual Data

all of the digital video compression systems currently under consideration. This includes the four systems proposed for broadcast Advanced Television and the six systems that have been under consideration for transmission of compressed 525-line signals by the group led by CableLabs. The DCT is also at the center of both the JPEG and MPEG systems we have mentioned previously. (Some of the other techniques being developed include Vector Quantization, Sub-Band Coding, Wavelets, and Fractals, but none of them is yet sufficiently developed for practical application in digital video compression applications.)

To understand the effect of the DCT, it helps to look at the results of applying it. Figure 10.2 shows the values of a typical 8x8 block of frame difference residual. Figure 10.3 shows the same data after transformation with the DCT. The values after the transformation process are called coefficients. Note how there are a few coefficients in the upper left hand corner that are of real significance, with virtually all the rest being low values, on the order of 1 or 2. It is this property of the DCT that makes it so useful for image compression – with the large proportion of nearly-zero values, the low values can be treated as zeroes, with little damage to the

235.6	-1.0	-12.1	-5.2	2.1	-1.7	-2.7	1.3
-22.6	-17.5	-6.2	-3.2	-2.9	-0.1	0.4	-1.2
-10.9	-9.3	-1.6	1.5	0.2	-0.9	-0.6	-0.1
-7.1	-1.9	0.2	1.5	0.9	-0.1	0.0	0.3
-0.6	-0.8	1.5	1.6	-0.1	-0.7	0.6	1.3
1.8	-0.2	1.6	-0.3	-0.8	1.5	1.0	-1.0
-1.3	-0.4	-0.3	-1.5	-0.5	1.7	1.1	-0.8
-2.6	1.6	-3.8	-1.8	1.9	1.2	-0.6	-0.4

Figure 10.3 – 8x8 Block of DCT Coefficients
After Transformation

reconstituted residual data when an Inverse DCT (IDCT) is applied at the decoder.

Following the DCT, there are many techniques that can be applied to further remove redundancy and thereby further compress the signal. These include quantization, variable length coding, Huffman coding, run length coding, perceptual weighting, and others. Some have an impact on the quality of the reconstituted frame difference residual while others are purely transparent mathematical manipulations. Different system designs make use of different combinations of these techniques, in varying orders and with different parameters, to yield the particular performance they are trying to achieve. In the end, it is fitting the best possible picture and sound through the limited channel bandwidth that is the determinant of the techniques used.

Unfortunately, in the initial publication, we were out of space at this point, so we had to wait a month to look into the meaning of the DCT coefficients and the further processing they undergo. Those of you reading this in book form can simply turn the page.

11

Crunching the DCT

In this installment, we will continue our examination of the DCT and the processing that follows it. Because of the complexity of these concepts, we will not have space both to deal with a non-technical subject and to get deeper into our technical discussion. We will try to resume inclusion of some of the non-technical issues in the near future.

DEVELOPING THE RESIDUE

At the end of last chapter's discussion, we had just begun to explore the concept of the Discrete Cosine Transform (DCT) as a means to reduce the redundancy in the residue of the motion compensation process. You will remember that motion compensation is a technique used to remove the field-to-field or frame-to-frame redundancy. It is used in combination with motion vectors that carry information through a side channel to the motion estimator in the receiver to help it predict motion in the current frame based on the previous frame it has held in memory.

After the motion compensaton process, we are left with a residue that represents either new information in the image that was not caused by motion or the leftovers that were not removed by the motion compensation because it is not a perfect process. That residue must be transmitted to the receiver so that it can reassemble a complete image by combining the residue with its prediction of the next field or frame. The mechanism for sending the residue is a major part of each digital compression system; we began looking at it last time with the DCT.

First published January, 1993. Number 7 in the series.

THE DISCRETE COSINE TRANSFORM

The DCT is a transform with the characteristic that it takes values in a matrix and concentrates most of the information about them into just a few coefficients. It produces just as many coefficients as there were matrix points to begin with, but the "energy" in the coefficients is mostly in one corner of the transformed matrix that results from the transform. This can be seen in Figures 10.2 and 10.3, which show a matrix of values and the DCT-transformed matrix, respectively.

The coefficients of the DCT can be thought of as representing various frequency bands within the original data block. The value in the upper left corner (where most of the energy is) is the dc or average level for the block. Continuing to the right from point-to-point in any row is increasing horizontal frequency. Continuing from top to bottom in any column is increasing vertical frequency. Thus the point in the lower right corner has both high horizontal and high vertical frequency. Not all of the bands are likely to have significant energy in them for any given picture, and advantage can be taken of this in reducing the amount of data to be transmitted.

CRUNCHING THE DCT

The DCT has several properties that make it a particularly apt choice for use in compressing digital video. Most of these relate to ways in which the values of the coefficients can be mathematically manipulated to reduce the amount of data to be transmitted. Most of this manipulation takes the form of number crunching.

First, the DCT coefficients are scaled using values in a *normalization* matrix that are known to both the encoder and the decoder. An example normalization matrix is shown in Figure 11.1. The normalization matrix has values associated with each point in the DCT array that control the effective step size of the quantization process to follow. It is like using an analog-to-digital converter with a variable number of bits to sample a signal. Instead of always representing the signal as 8 or 10 bits, for example, one sample can be 8.2 bits, the next one 5.4 bits, and so on.

The scaling for each point in the DCT array is thus determined by the associated value in the normalization matrix, and the normalization matrix can be made selectable or updateable at both the encode and decode ends on a dynamic basis. Varying the normalization matrix allows control of the quality and bit rate of the encoded image. This becomes very important when considering rate control into the output buffer.

16	11	10	16	24	40	51	61
12	12	14	19	26	58	60	55
14	13	16	24	40	57	69	56
14	17	22	29	51	87	80	62
18	22	37	56	68	109	103	77
24	35	55	64	81	104	113	92
49	64	78	87	103	121	120	101
72	92	95	98	112	100	103	99

Figure 11.1 – Example Normalization Matrix

15	0	-1	0	0	0	0	0
-2	-1	0	0	0	0	0	0
-1	-1	0	0	0	0	0	0
0	0	0	0	0	0	0	0
0	0	0	0	0	0	0	0
0	0	0	0	0	0	0	0
0	0	0	0	0	0	0	0
0	0	0	0	0	0	0	0

Figure 11.2 – Normalized and Quantized DCT Coefficients

240	0	-10	0	0	0	0	0
-24	-12	0	0	0	0	0	0
-14	-13	0	0	0	0	0	0
0	0	0	0	0	0	0	0
0	0	0	0	0	0	0	0
0	0	0	0	0	0	0	0
0	0	0	0	0	0	0	0
0	0	0	0	0	0	0	0

Figure 11.3 – Denormalized DCT Coefficients

144	146	149	152	154	156	156	156
148	150	152	154	156	156	156	156
155	156	157	158	158	157	156	155
160	161	161	162	161	159	157	155
163	163	164	163	162	160	158	156
163	163	164	164	162	160	158	157
160	161	162	162	162	161	159	158
158	159	161	161	162	161	159	158

Figure 11.4 – Reconstructed 8x8 Block of Frame Difference Residual Data

We will be discussing this in the next chapter.

Following normalization, the coefficients are *quantized*. Quantizing is essentially rounding off the coefficients to the nearest integer. As can be seen in Figure 11.2, the lowest values in the DCT matrix of Figure 10.3 have been reduced to zeroes by the combination of normalization and quantization. This can be thought of as selecting only the significant coefficients for transmission. At the decoder, the DCT coefficients are *de-normalized* by multiplying by the values of the normalization matrix, the results of which are shown in Figure 11.3. When all of this is done, there will be some distortion when the DCT is ultimately reversed in the decoder through application of the Inverse DCT (IDCT), but the distortions will be small if the process is applied cleverly. This can be seen in the reconstructed residual data block shown in Figure 11.4. Here the transmitted data has been de-normalized and the IDCT applied. The amount of error caused to the original frame difference residual data can be seen in Figure 11.5, which shows the differences between the original values in Figure 10.2, before application of the DCT plus quantization and normalization, and the recovered values of Figure 11.4, after de-normalization and application of the IDCT.

-5	-2	0	1	1	-1	-1	-1
-4	1	1	2	3	0	0	0
-5	-1	3	5	0	-1	0	1
-1	0	1	-2	-1	0	2	4
-4	-3	-3	-1	0	-5	-3	-1
-2	-2	-3	-3	-2	-3	-1	0
2	1	-1	1	0	-4	-2	-1
4	3	0	0	1	-3	-1	0

Figure 11.5 – Error: Difference of Original and Reconstructed 8x8 Block

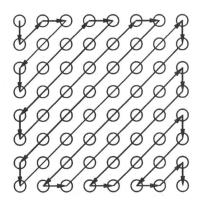

Figure 11.6 – Zig-Zag Scanning of DCT Coefficients

ZIG-ZAG SCANNING

Another technique that can be applied to the DCT is the *zig-zag scanning* of the coefficients. By following a zig-zag order through the coefficient matrix when transmitting the coefficients, it is possible to create long runs of successive zeroes. One possible zig-zag pattern is shown in Figure 11.6. With long zero runs, *run length coding* can be applied to indicate how many zeroes come in succession, rather than actually sending all the zeroes. This results in less data being required to send the same information.

VARIABLE LENGTH CODING

Now that we have normalized and quantized the coefficients, zig-zag scanned the coefficient matrix, selected the significant coefficients, and run length coded the others, we must find an efficient way to send the data that results from all this manipulation. One method for doing this recognizes that all of the codes to be sent do not have equal probability of occuring in the data. This allows a form of coding to be devised that minimizes the overall number of bits that are required to send any particular set of data values. The method is *variable length coding*.

The value of variable length coding can be seen through use of an example. In Table 11.1, there is a source of data, S, with four different possible symbols, s_1, s_2, s_3, s_4. Each of the symbols has a probability associated with it. Two different codes are shown. Code I uses fixed length coding. Since there are four symbols, two bits can be

Table 11.1 – Example of Symbol Coding

Symbol	Probability	Code I	Code II
s1	0.60	00	0
s2	0.30	01	10
s3	0.05	10	110
s4	0.05	11	111

used to represent them. Code II uses variable length coding. The symbols are represented by one, two, and three bits, depending on their probabilities.

The result of these codings is that Code I has an average length of 2.0 bits per symbol, while Code II has an average length of 1.5 bits per symbol. This is the result of multiplying the number of bits of the code word associated with each symbol by the symbol's probability and then summing the results. This is a savings of 25 per cent in the amount of data required to carry these symbols by using the variable length Code II instead of the straightforward Code I. Nothing comes for free, however, and the use of a variable length code adds a requirement for buffering and rate control in order to fit into the constant rate of the RF transmission channel. This is part of what we will explore in the next chapter.

You might be interested to know that Code II is what is called a prefix condition code. This means that no codeword is a prefix of any other codeword. This permits a serial bit stream of Code II codewords to be uniquely decodable. A method for generating what is called a "compact code" of variable length was developed by Huffman. This is what is referenced when systems are described as using Huffman code. An example of a system using Huffman coding is the Group III fax that is nearly ubiquitous today.

Well, that's it for this time. In the next chapter, we will look at the impact that using variable length coding has on the system, as it is rather far-reaching. We will also look at some of the black magic that goes into digital video compression systems in the form of perceptual weighting and other esoteric techniques.

12

The Buffer and the Human Visual System

Before we launch back into our continuing technical discussion of digital video compression, it is about time that we look at one of the most important **management** decisions that will have to be made regarding the technology to be used in each television operation that uses digital video compression. As suggested in the past, those of you who are technical readers may wish to call the attention of your nontechnical associates to this discussion.

THE "LOOK" OF PROGRAM CONTINUITY

Every television program service, whether it is a broadcast network, a cable network, a television station, a local cable operator, a direct broadcast satellite operation, or an industrial programming operation, has a certain "look" to its program continuity. That look consists of the way in which it makes transitions from one program to another. It includes such factors as whether there are few or many bridge elements in the program stream, whether there is promotional material, whether and how identification is done, and whether there are any "effects" at transitions from one program element to another. Managers of such operations are often quite concerned with the look of their service because it can have a substantial impact upon how its viewers perceive their service.

The coming of Advanced Television, be it broadcast HDTV or Compressed 525-line systems for cable and DBS, will lead to choices between cost and capability that will have extensive and long term consequences for the operations

First published February, 1993. Number 8 in the series.

implementing ATV. The likely choices will be between minimum cost coupled with a reduced look to the program continuity or considerably greater cost while maintaining the production values incorporated in today's program continuity.

Much of the choice will revolve around where encoders (compressors) are located in the total system. Since they will be among the most expensive equipment for ATV, at least at the start, one of the ways to reduce system cost will be to minimize their number. Putting the encoders as early in the production/distribution chain as possible and distributing fully compressed signals is a method to accomplish this.

CENTRALIZED ENCODING AND PASS-THROUGH

Consider the distribution of a local commercial, for instance. If the production house sends copies to each of the local television stations and to each of the local cable operations in an uncompressed form (e.g. in components on Beta or M-II tape, in NTSC on U-Matic or S-VHS tape, in digital form on D-2 or D-3, or on analog or digital HDTV tape), each station or cable operation will require an encoder to process and air the commercial.

On the other hand, if the production house has an encoder, it can send compressed copies of the commercial to all of the stations and headends. This way, while all of the various operations will require a tape machine that will handle the compressed material, they can avoid the much larger cost of an encoder. Expectations are that tape machines will be fairly inexpensive compared to encoders, since they can be based on the consumer recorders that should be available from the start of ATV. In essence, the production house provides a shared encoder for all of the operations, and the total cost to all operations is minimized.

The same idea applies to network distribution. The encoder can be located in one of two places: Each station or headend can have one, in which case each must bear the high cost of ownership, and the network would feed uncompressed or partially compressed signals. Or, the network can have one, delivering fully compressed signals to each station or headend. The latter arrangement is called "pass-through" because each station or cable operation passes through the signal from the network with no further compression required.

CUTS ONLY

But there is a down-side to this approach. Once a signal has been compressed, it can only be cut into, unless it is decoded and later re-encoded. Re-encoding requires an encoder, which is the cost that was intended to be avoided by the approach. Cuts-only has some serious implications. It means, for example, that instead of appearing as a lower-third super crawl, an emergency message or news bulletin would have to be inserted full-screen, interrupting the program material into which it was inserted. Similarly, video tags on the ends of commercials would have to be full-screen, instead of appearing as supers over background video from the spots.

Because of the technical structure of the compressed signals, simply cutting or switching between two pieces of material in compressed form, a program and a commercial, for example, is not straightforward. In order to make a clean transition from one to the other, the outgoing material can only be exited at certain times related to the way the signal is processed. The incoming material can only be joined at specific times as well. Once a decision is made to switch, it will be necessary to wait for the next occurrence of the appropriate points, if a clean switch is to occur. Alternatively, a switch can be made at an arbitrary point, but it may look like a "channel change" depending on the particular system. A channel change means different things for the different proposed systems; a mosaic wipe is a good model for imagining the effect with several systems.

Making the switch at black is the easiest to accomplish for most ATV systems. This could require that the material have prerecorded fades to black on the outgoing ends and fades up from black on the incoming ends of program segments and other elements. It is also possible that the segments will require special markers in the data to indicate where switches can be made. To allow switching at designated points, such markers might be applied automatically by the encoder or they might be forced by the operator making the encoded recording.

ENCODING AT THE END

The alternative to these limitations is to do the switching and effects in a non-compressed form. This means that an encoder must come at the end of the distribution chain, after all of the switching and effects have taken place. This puts an encoder at the output of a broadcast studio, prior to the transmitter, or at the output of a cable commercial insertion system, prior to the modulator and the trunk distribution network.

The distribution of material and its recording and playback might be done with some level of compression, in which case it will have to be decoded before post processing. This can mean a cascade of compression and decompression, depending on the system design. Compression and decompression ultimately reduce image quality, and care will be required in system designs to maintain the best quality. Small amounts of compression can be applied with little degradation and at moderate cost.

The use of non-compressed signals downstream, in the local operations, can provide all of the production values associated with program continuity today. One trade-off for achieving this is the cost of one or more encoders per channel of output. In a large cable system, with a current cost for encoders on the order of $100,000 each, this can be prohibitive. Encoders for broadcast ATV (at the HDTV level) are expected to cost roughly double this amount at the start. Another trade-off is the potential loss of image quality if multiple compressions and decompressions are applied.

All of this says that managers of television operations are going to have to decide what their interests are in terms of production values in their program continuity and how much they are willing to pay to achieve those interests.

Compromises may be required. Careful system design, with an eye to what will be needed in the future as the audience for ATV grows and the "look" of the ATV program continuity becomes more important, is an absolute necessity.

CLOSING THE LOOP

Turning now to the technical side of our discussion, we ended the last chapter with a discussion of variable length coding and the efficiency that could be achieved through its use. In the simplified example we examined, we saw a reduction in the average amount of data required from 2 bits-per-data-symbol with fixed length coding to 1.5 bits-per-data-symbol with variable length coding. This is a very substantial reduction of 25 per cent in the amount of data that must be transmitted.

This improvement in efficiency comes at a price, however. The price is that the coding is *variable length*. This means that it will take differing amounts of time to send a given number of data symbols, depending on the contents of those symbols, if the bit rate is held constant. Looked at another way, to transmit a constant symbol rate, as would be required to send a fixed amount of data about each field or frame of a television image, the bit rate to be transmitted would have to vary. This doesn't work because the channel coding requires a fixed bit rate in order to allow proper recovery of the carrier and the clock in the receiver.

Further complicating this situation is the fact that most digital video compression systems are adaptive, i.e. they change the way they process the image depending upon the image content. This produces even more variability in the data to be transmitted than would be caused by variable length coding alone.

THE BUFFER AND FEEDBACK

What to do? The answer is a buffer with feedback to control some parameter of the compression process in order to smooth out the bit rate that must be transmitted. (See Figure 12.1.) The feedback is based on the *buffer fullness*, a measure of how much data is in the buffer at any instant waiting for transmission. As the buffer empties, the feedback will allow more bits to be generated, thus filling the buffer to the target fullness level. As the buffer fills up, the feedback will cause a reduction in the number of bits generated, bringing the buffer fullness back to the desired level.

What to control? In last chapter's discussion, we described the *normalization matrix* as a mechanism for determining how many bits are used to represent each coefficient in the transformed matrix following a Discrete Cosine Transform (DCT) of the motion-compensation residue of a pixel block. Suppose, instead of one normalization matrix, we had a wide assortment of normalization matrices at our disposal. They could produce larger and smaller numbers of bits per coefficient. By selecting among the normalization matrices based on buffer fullness, we could control the amount of data going into the buffer so that, on average, it matched the amount of data we took out of the buffer at a fixed rate for transmission. This, in essence, is how the variable data rate of the variable length coding is compensated in order to present a fixed data rate to the channel coding system.

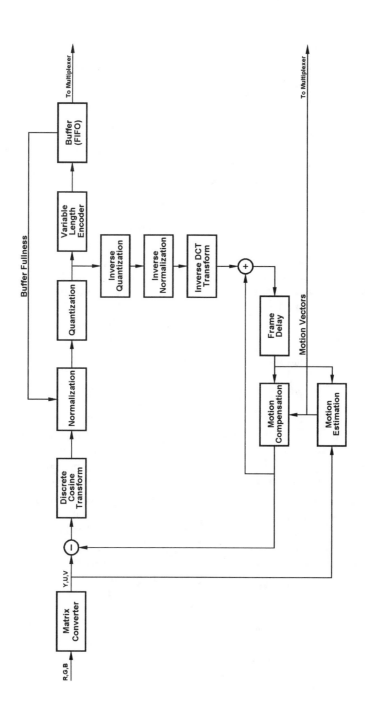

Figure 12.1 – Generic Digital Video Compression Encoder (Simplified)

The use of a buffer and the selectable normalization matrix provides the greatest opportunity for differentiation of the designs of the several systems. Normalization matrices can be designed to favor various aspects of the image that determine the quality of the picture produced at the receiver. When there is a lot of motion in the image combined with image complexity, the amount of residue to be transmitted tends to increase tremendously. When there is a still image with low complexity, there is not so much to send.

HIDING BEHIND THE HUMAN VISUAL SYSTEM

As the amount to be sent goes up, the accuracy (number of bits) that can be allocated to each coefficient goes down. This increases the distortions in the picture. Fortunately, the human visual system is not so sensitive to distortions in moving images or ones of great complexity. This provides perceptual masking of the distortions generated.

A great deal of intelligence is devoted to selection of the normalization matrix. Duplicate copies are kept in the encoder and the decoder, and instructions are sent in a side channel from the encoder to tell the decoder which choice to use in de-normalizing the coefficients. Because of the frequency relationship of the DCT coefficients, the various coefficients can be weighted in the normalization matrix to optimize for different qualities in the original image. (See Figure 12.2. In the last chapter, we described how the coefficients in the DCT-transformed matrix represent increasing horizontal frequency across the matrix and increasing vertical frequency down the matrix, starting from the dc value in the upper left hand corner.) Factors considered might include noise, resolution, image detail, dynamic range, and motion and color perception of the human visual system.

Another purpose of the selections made between normalization matrices can be the support of *statistical multiplexing*. This is an approach that can be used when multiple compressed signals are to be combined into a single transmission channel. Instead of allocating a

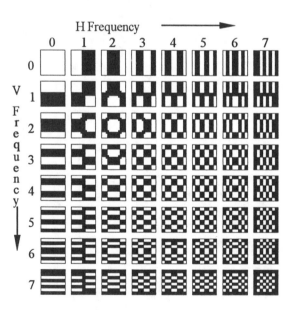

Figure 12.2 – Patterns Represented by
Coefficients of an 8x8 Discrete Cosine
Transform (DCT)

fixed number of bits in the transmission channel to each of the several compressed signals to be combined, they can be allocated variable numbers of bits depending upon the complexity of the images they are carrying. This requires a means of balancing the data from all the channels so that the average total data into the buffer equals the amount taken out at a fixed rate. The buffer fullness must therefore simultaneously control the normalization matrices of all the channels being combined into the buffer.

If all goes according to the current plan, by the time the next issue of *TV Technology* reaches you, there will be a decision by the FCC Advisory Committee on the system to recommend for terrestrial broadcast ATV. If that happens, we will try to have a detailed description of the selected system for you in next month's installment.[11]

---

[11] The Advisory Committee's Special Panel met for most of a week in February, 1993, but the results were not as had been hoped. See Chapter 2 for a description of what happened and how it led both to the certainty of a digital system and the formation of the Grand Alliance

_____ PART 4

There's Audio in Advanced Television, Too

Second only in importance to digital video compression as an enabler of Advanced Television, digital audio compression does not require such high compression ratios as its video cousin. But it must work just as hard because of the human aural system's greater sensitivity to audio errors than the human visual system's rather forgiving response to video errors. Put another way, the human visual system lets us do a bunch of dastardly deeds to the video while still recognizing it as a pretty good replica of that with which we started. The human aural system, on the other hand, is rather unforgiving, catching us at the least discrepancy in the reproduced sound, within certain bounds. The result is that we can get away with something like 30-50-to-1 compression of video while audio runs at somewhere between 4-to-1 and 8-to-1 for high fidelity (e.g. music) applications.

There are two principal audio compression systems currently being used in Advanced Television applications. They are the MPEG audio system, as originally developed for MPEG-1 and extended for MPEG-2, and the Dolby AC-3 system. Each gets its own chapter in this section.

AC-3 has recently been selected for use as part of the Digital Video Disk (DVD) lineup of techniques, together with MPEG-2 video. But in keeping with the continuing state of confusion in the selection of audio standards that is discussed in Chapter 13, it was only selected for use on "NTSC-compatible" (i.e. 59.94 Hz) DVDs, while MPEG audio was selected for use on "PAL-compatible" DVDs (50 Hz versions). To further confuse matters, the computer industry is currently pushing the Hollywood studios to produce DVDs with MPEG audio or uncompressed audio in addition to AC-3. Although it might, don't count on it happening, however, because if they were made that way, there would be little room left for the video on normal-capacity DVDs.

The bottom line to all this is — don't forget, "There's Audio in Advanced Television, Too!"

13

MPEG Audio Coding

To many people in the television business, television has mostly to do with images and video. To others, television is "radio with pictures." In truth, television is an experience founded on an amalgam of images and sound, video technology and audio technology. So far in this column, we have dealt with video coding, transport systems, modulation schemes, and similar topics, but not audio. We are now going to correct that omission by taking a couple of installments to look at the state of the technology for audio compression and encoding for Advanced Television.

FIRST, A LITTLE HISTORY...

There currently is no single technique or standard to which one can point and say that it is the audio method for advanced television. This derives in many ways from the history of the MPEG and ACATS activities over the past half dozen or so years. In each case, tests of competing audio systems have been conducted using subjective assessment techniques, with not altogether consistent results, and selections have been made. Time has shown, however, that the choices made have not turned out to be final, at least not yet.

The first of the competitive situations was the MPEG-1 audio system selection. A number of consortia put forward several different approaches to audio compression coding. The one selected was called Musicam, and it now is enshrined within the MPEG-1 audio coding system. MPEG-1 encodes a single stereo pair of channels.

First published February, 1995. Number 26 in the series.

MPEG-1 audio coding was used with one of the proposed Advanced Television systems in the FCC Advisory Committee selection process, but the other systems each had a different associated method. When the Grand Alliance formed, there was no agreement among the participants as to which audio system to select, so the Advisory Committee and the GA conducted a side-by-side comparison of several systems. The Dolby AC-3 system won, with MPEG-1 audio the runner-up. Consequently, the Grand Alliance system now incorporates AC-3.

Later, the MPEG-2 system development effort needed an audio encoding method. Following recommendations from SMPTE, EBU, and others, it was decided to increase the audio capacity to five channels to include a center channel and left and right surround channels in addition to the normal left and right channels. Provision was also made for a low frequency effects channel (called LFE – essentially a subwoofer) that had much lower bandwidth than the other channels. The complete system was therefore referred to as a 5.1-channel system.

There was a desire in the MPEG-2 effort to incorporate an audio system that was backwardly compatible with the system of MPEG-1. The MPEG-1 method was extended to include capability for 5.1 channels in a manner that allowed older MPEG-1 decoders to extract the basic stereo pairs while newer decoders could recover the full complement. It is also true that extended MPEG decoders can recover MPEG-1 compliant audio, making them forwardly compatible as well.

At the same time, there were assertions being made by a number of organizations that some of the newer coding techniques could provide far better performance than MPEG audio, so more tests were held. Some of the systems did, in fact, exceed the performance of MPEG audio, among them Dolby AC-3 and a system from AT&T. This led to a movement to permit another audio technique to be used as part of MPEG-2 coding, this variety to be Not Backwardly Compatible (NBC – not to be confused with the network), i.e. it will work only with new decoders. The current situation is that MPEG-2 specifies the backwardly compatible extended MPEG-1 audio and work is continuing to select and document an NBC audio system for inclusion as part of MPEG-2 some time in the future.

Use of the various audio systems in current and planned applications depends largely upon who is making the selection, when it is being made, and the past associations between some of the parties. For instance, MPEG audio is being used by DIRECTV in its Digital Satellite System (DSS). The DSS decoders are first being built by Thomson Consumer Electronics. Thomson, along with Philips, Matsushita, and others developed Musicam. Dolby AC-3 was first used in the FCC Advisory Committee activity by General Instrument. GI is now incorporating AC-3 into the DigiCipher-II set top converters that it is planning to sell to the cable and wireless cable industries. Other examples abound.

Given the prevalence of MPEG audio and AC-3 in current applications and plans, we will examine both in a little detail. We will start with MPEG audio this time and finish with AC-3 next time.

HUMAN AUDITORY SYSTEM

Before getting into the details of audio compression systems, however, it is important to understand a little bit about how the human auditory system works. This background is necessary to understanding how the compression systems work since both are based on the same characteristics of the human auditory system, although the approaches taken differ somewhat.

The current fundamental understanding of the operation of the ear/brain combination is that they act as a multi-channel real time analyzer with varying sensitivities and bandwidths through the audio range. This means that, if a tone exists within a certain frequency range, any other sounds (tones, noise, or a combination) that are lower in level than that tone and within a certain frequency range of that tone will be "masked" and not heard by the listener.

Experiments have shown that a critical bandwidth exists at any point in the audio range. To be heard, a lower level signal must fall outside that bandwidth relative to a higher level tone. There are 24 such critical bands in the range from 20 Hz to 15 kHz. It is important to note that the higher in level the masking tone, the wider is the bandwidth, particularly on the high frequency side, in which other sounds will be masked. On the low frequency side of the masking tone, the masking effect tends to fall off quite rapidly. It also happens that the greatest amount of masking occurs in the region closest in frequency to that of the masking signal. For high sound levels in the mid-range frequencies, there is also a significant masking effect at the second harmonic of the masking tone.

Use of single tone masking is an oversimplification, of course, since most program material, especially music, comprises a complex combination of sounds. Since the job of the audio coder is to represent all the sounds in the source without any audible distortion while using about one-sixth of the amount of data required for the uncoded digital audio input signal, allocations of the bits available in the coded output must be made between the sounds arriving at the input. How this is done is part of what differentiates the systems from one another.

Another factor in the workings of the human auditory system is what is termed stereophonic irrelevance. This is a recognition that there is information that is captured in a stereophonic pickup of a sound field that is irrelevant to the spatial perception of the stereophonic presentation. These stereo-irrelevant signal components are not masked, but they do not contribute to the localization of sound sources. They are ignored in the binaural processing of the human auditory system. This means that stereo-irrelevant components from any direction in the sound field can be reproduced through any loudspeaker or combination of loudspeakers without affecting the stereophonic impression.

It should be noted that the impact of the various encoding techniques on perceived sound is quite subjective and depends heavily on the material being coded. Objective measurements will tell little that can be used to differentiate the systems. When necessary, statistical subjective assessment is required using listeners with good listening skills to identify the superior of a group of systems. It should also be noted that the differences between systems of this caliber are likely to be apparent only on high quality audio presentation systems.

MPEG AUDIO

In the basic MPEG audio system, the input signal is broken up into uniform time segments, called "frames," based upon a certain number of samples. The frames of audio data are then filtered in a polyphase filterbank to separate them into sub-bands of audio energy contained within each frame. A polyphase filterbank is a set of equal bandwidth filters with special phase interrelationships that allow for an efficient implementation of the filterbank. It is normally implemented through a transform process rather than with actual hardware digital filters.

Since the filtering is done digitally, with clocking at the sample rate or a multiple thereof, the overall audio bandwidth and the bandwidth of each sub-band will depend upon the sample rate in use. A number of different sample rates are provided, the exact choices depending upon the version. Thus the bandwidths of the individual bands automatically follow the overall bandwidth as determined by the sample rate.

The sub-band sub-samples, which constitute a frequency domain representation of the original time domain input, are quantized by means of a quantization scale factor similar to those used for video coders and which we have described previously. The quantization performs essentially a companding function, compressing the dynamic range of the audio in each sub-band. This results in a set of coefficients for each sub-band for each frame period.

Simultaneously with the calculation of the sub-band coefficients, a frequency-dependent masking threshold is calculated for each frame. This masking threshold is based on a psycho-acoustic perceptual model and alters the bit allocation process for each sub-band. After the sub-band sub-samples have been generated and quantized, bits in the encoder output are dynamically allocated to each sub-band, subject to the ratio of the signal within a sub-band and its masking function.

There are layers of processing provided in MPEG audio, with Layer I yielding all of the basic functionality and Layer II adding more sophisticated processing in deriving the quantization scale factor and additional coding of the bit allocation. Layer III provides for additional decomposition of the frequency components in each sub-band through the use of a modified discrete cosine transform (MDCT).

Following development of the coefficients, various forms of data reduction are applied similar to those we have described on prior occasions for the video system. An example is variable length coding, a mathematically fully reversible procedure in which shorter code words are assigned to frequent events and longer code words are assigned to less freqeuent events. These data reduction techniques take advantage of the characteristics of the human auditory system wherever possible.

At the MPEG audio decoder, the process is essentially reversed, with the bits for each sub-band de-quantized and applied to a synthesis filterbank that reconstructs a PCM audio signal from the sub-band samples sent to it.

MPEG-1

The MPEG-1 audio system provides two channels of stereo audio at sampling rates of 32 kHz, 44.1 kHz, and 48 kHz. To help clarify our examination, we will consider only the 48 kHz rate. All of the principal parameters of MPEG audio in both its original MPEG-1 form and its extended MPEG-2 form are shown in Table 13.1.

The frames used in Layers II & III are 1152 samples long. There are 32 sub-bands, and the sampling frequency at the output of each filter is the sampling frequency (Fs) divided by 32 (Fs/32). Since the Nyquist criterion

Table 13.1 – MPEG-2 Backwardly Compatible Audio Coding System

Characteristic	Sampling Frequency					
	16 kHz	22.05 kHz	24 kHz	32 kHz	44.1 kHz	48 kHz
Audio Bandwidth	7.5 kHz	10.3 kHz	11.25 kHz	15 kHz	20.6 kHz	22.5 kHz
Frame duration - Layer 1	24 ms	17.4 ms	16 ms	12 ms	8.7 ms	8 ms
Frame duration - Layer 2	72 ms	52.5 ms	48 ms	36 ms	26.25 ms	24 ms
Frame duration - Layer 3	72 ms	52.5 ms	48 ms	36 ms	26.25 ms	24 ms
Number of Sub-bands	32	32	32	32	32	32
Sub-band Sampling Freq. (Fs/32)	500 Hz	689.0625 Hz	750 Hz	1 kHz	1378.125 Hz	1.5 kHz
Sub-band Bandwidth (Fs/64)	250 Hz	344.5313 Hz	375 Hz	500 Hz	689.0625 Hz	750 Hz
Bit Rate - Layer 2/Stereo	96 kb/s	128 kb/s	128 kb/s	192 kb/s	256 kb/s	256 kb/s
Bit Rate - Layer 2/5.1 Channel	128 kb/s	192 kb/s	192 kb/s	256 kb/s	384/kb/s	384/kb/s
Coding Modes	3/2, 3/1, 3/0, 2/2, 2/1, 2/0, 1/0 second stereo program up to 8 additional multi-lingual or commentary channels associated services					

holds that the highest frequency that can be represented in a sampled system is half the sample rate, the bandwidth of the sub-bands is Fs/64. Looking at these relationships again in terms of 48 kHz sampling, the filters have output sample rates of 1,500 Hz and bandwidths of 750 Hz with uniformly sharp skirts.

At 48 kHz, the frames are each 24 milliseconds long. This results in 36 sub-samples of the input signal falling in each of the 32 equally-spaced sub-bands that extend up to the 24 kHz Nyquist frequency implied by 48 kHz sampling. Of course, the actual audio bandwidth supported is somewhat less than the Nyquist frequency to allow for practical filter implementation, e.g. yielding 22.5 kHz audio bandwidth with 48 kHz sampling.

The MPEG-1 version of the MPEG audio system provides for what is termed joint stereo coding, which is a method for exploiting stereophonic irrelevance. A particular form of joint stereo coding included is called intensity stereo coding; it removes redundancy in stereo audio information by retaining only the energy envelope of the right and left channels at high frequencies.

The basic MPEG-1 audio frame, after encoding, includes four basic types of information. These are a header, a cyclic redundancy code (CRC) for error checking, audio data, and ancillary data. The audio data consists of bit allocation information, scalefactor selection information, scalefactors, and sub-band samples (the largest part of the data). The length and usage of the ancillary data are not specified to allow the widest number of possible applications to take advantage of this capability.

MPEG-2

The MPEG-2 audio coding system is an extension of the system adopted for MPEG-1 and is both forward and backward compatible with it. The MPEG-2 extensions add three additional sampling frequencies and add up to four more channels, permitting 5-channel surround sound with support for low frequency effects. The MPEG-2 system can deliver "Compact Disk quality" stereo sound in 256 kilobits/second (kb/s) and 5.1 channel surround with LFE in 384 kb/s.

Besides the capability for 5.1-channel surround sound, the MPEG-2 version also enables the use of the audio system for independent carriage of separate material in several different configurations within the same audio elementary stream. This could be used in cases where multi-lingual transmission of a program is necessary or to include clean dialogue for the hard-of-hearing or a commentary channel for the visually impaired. In order to provide all of these capabilities, a variety of encoder input and decoder output modes are identified.

The MPEG-2 extensions achieve their compatibility with MPEG-1 decoders through use of the ancillary data field. All of the additional data beyond that used in MPEG-1 is carried as ancillary data. Since an MPEG-1 decoder basically ignores this field, it continues to extract the stereo information for which it was designed. An MPEG-2 decoder, of course, is designed to use all of the information and can extract the full range of services available.

DOLBY AC-3

As mentioned earlier, Dolby AC-3 is the other important audio technique receiving wide application. We will take a look at it in the next chapter.

14

Dolby AC-3 Audio Coding

With the recognition that high quality audio can dramatically improve the impact of the television viewing experience, audio has taken on increased importance in the development of advanced television. Our last time together, we began an examination of the audio compression technology that will be used as part of advanced television systems. We started with the MPEG audio system developed for MPEG-1 and extended for use with MPEG-2 in a forwardly and backwardly compatible manner. This time, we will look at one of the non-backwardly-compatible audio systems that will be used in advanced television – Dolby AC-3.

A COMPARISON

In order to differentiate the AC-3 system from the MPEG audio system we discussed in detail last time, it will help to briefly review the MPEG audio encoding techniques as a point of departure for our comparison. This will also give us the opportunity to look at MPEG audio compression from a systems level perspective.

The MPEG audio system encoder receives input samples of the audio signal and divides them up into frames 1152 samples, or 24 ms, long (for Layer II & III systems). As shown in Figure 14.1, the samples are sent to a filter bank and to a bit allocator. The filter bank divides the spectrum of each frame into 32 sub-bands, each 750 Hz wide. The filter outputs are 36 sub-samples per sub-band for each frame.

First published March, 1995. Number 27 in the series.

Encoder

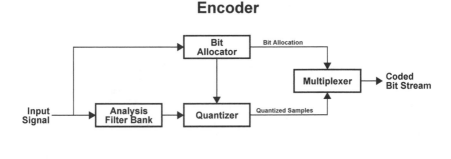

Decoder

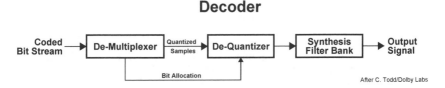

After C. Todd/Dolby Labs

Figure 14.1 – Forward Bit Allocation Encoder & Decoder

The importance of the data in each sub-band to the re-creation of the sound is determined in a bit allocator. The amount of data in the coded bit stream then is controlled and compression is achieved by selecting sub-bands to be sent to the decoder and by quantizing the sub-sample data, that is by increasing the granularity of the data (reducing its amplitude resolution) according to the bit allocation.

Following the bit allocation and quantization processes, the bit allocation values and the quantized coefficents are multiplexed and sent to the decoder. This makes MPEG audio a Forward Adaptive Bit Allocation system, which means that bit allocation decisions are made adaptively, based on signal content, and that the decisions are sent forward from the encoder to the decoder so that the decoder can properly de-quantize the quantized values sent to it.

If there is a forward adaptive bit allocation method, there must be a Backward Adaptive Bit Allocation scheme, and it is shown in Figure 14.2. In this case, the input sample frames are sent just to the filter bank. The filter bank produces a number of output samples in each of its sub-bands for each sample frame. The output samples are the coefficients that are processed for transmission to the decoder.

Each coefficient is structured as a combination of an exponent and a mantissa. The exponent carries the major level information for its sub-band, and the mantissa carries the detailed level of each sample. Because the exponent is exponential, it can be thought of as carrying the amplitude "range" information for the sub-band. The use of exponents permits a wide dynamic range to be carried.

Encoder

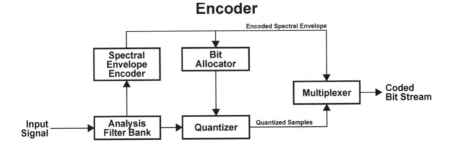

Decoder

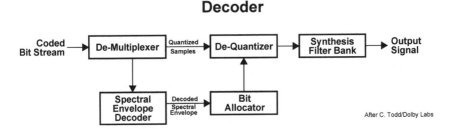

After C. Todd/Dolby Labs

Figure 14.2 – Backward Adaptive Bit Allocation Encoder & Decoder

In the backward adaptive bit allocation method, the collection of exponents at a given sample time forms an overall representation of the signal spectrum that is called the spectral envelope. The spectral envelope is encoded and sent to the decoder. It is also used to decide which coefficients are significant to the sound at any instant and then to control the bit allocation to each of the sub-band coefficients.

The bit allocation is realized by quantizing the coefficients for each sub-band according to the calculated allocation. The quantization is similar to that used in MPEG audio and controls the number of bits used to represent a value by changing the granularity, or the "fineness" of amplitude resolution, with which the value is expressed.

The quantization process leads to differences between the real values and the values used to represent the coefficients. These differences constitute "quantization noise" that tend to decrease the signal-to-noise ratio of the coded sound. Because the reconstruction filter, called the "synthesis filter bank," used in the decoder limits the quantization noise in any sub-band to be nearly the same frequency as the signal in that sub-band, the quantization noise is "masked" by the signal, yielding a higher apparent signal-to-noise ratio. We'll look at masking in more detail momentarily.

Once the spectral envelope has been encoded and the mantissas of the coefficients have been quantized, the results of the two processes are multiplexed together to form the coded bit stream. The coded bit stream, in

Encoder

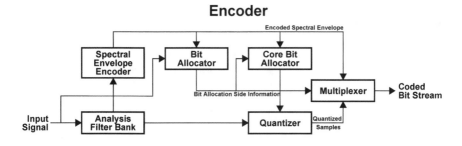

Decoder

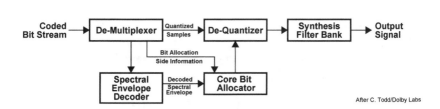

After C. Todd/Dolby Labs

Figure 14.3 – Hybrid Backward/Forward Adaptive Bit Allocation
Encoder & Decoder

turn, is packetized for multiplexing with other types of bit streams such as video and external data and sent to the decoder.

At the decoder, the spectral envelope is used once again to calculate the bit allocation. Since the same spectral envelope is used in both encoder and decoder, the exact same bit allocation decisions will result at both ends. This allows the decoder to de-quantize the quantized mantissas without the actual bit allocation data being sent. The result is a smaller amount of data to be transmitted, with all of the data devoted to coded audio. The fact that the content-controlled bit allocation is calculated identically at both ends is what makes this a backward adaptive bit allocation process.

AC-3 ENCODING

The Dolby AC-3 system uses a hybrid of forward and backward adaptive bit allocation. The basic system arrangement is shown in Figure 14.3. Added to the straight backward adaptive bit allocation system is a core backward adaptive bit allocation routine that runs identically in both the encoder and the decoder. The core routine can be modified in both places by forward adaptive bit allocation side information. Thus both the encoder and decoder core routines run independently using the spectral envelope, as in the backward adaptive case, but they are altered by forward adaptive information that improves their accuracy.

The core bit allocation routine is relatively simple and is based on a fixed psychoacoustic model that makes certain assumptions about how acoustic masking takes place in the human auditory system. The forward adaptation then makes two types of modifications: to the parameters of the psychoacoustic model and as differences to the bit allocations that would result from the current psychoacoustic model.

ADVANTAGES AND DISADVANTAGES

The forward adaptive bit allocation system has the advantages that the psychoacoustic model resides only in the encoder and that it can be improved over time as better models of the human auditory system are developed. Theoretically, it can be as accurate as desired. Since explicit bit allocation information is sent to the decoder, previously installed decoders will track changes in the model as they are made. The downside to all this is that a substantial amount of the data capacity devoted to each audio channel must be absorbed for transporting the bit allocation data. For example, the MPEG audio Layer II encoder generates almost 4 kbits/second/channel when operated at 48 kHz sampling, with 24 msec time resolution and 750 Hz frequency resolution.

Depending upon the content of the audio material being coded at any instant, it is desirable to alter the balance between the frequency resolution and the time resolution. When transients are significant, it would be better to update the bit allocation with finer time resolution, but with forward adaptive bit allocation, this results in a direct increase in the side channel information data rate. Our 4 kb/s example cited in the last paragraph, for instance, would become 12 kb/s if the time resolution were increased to 8 msec.

Similarly, during steady state conditions, it would be better to provide finer frequency resolution in the spectral analysis and in the allocation of bits. In our MPEG audio Layer II example, it is quite possible that virtually every 750 Hz frequency band could contain significant information. In such a case, it would be possible to obtain only limited bit rate reduction before incurring audio degradation. But increasing the frequency resolution by 8 times, for example, would increase the side channel bit allocation data by the same 8 times.

The backward adaptive bit allocation method overcomes the disadvantages of the forward adaptive method, but it has disadvantages of its own. Since both the encoder and decoder use the same information (derived from the spectral envelope) to derive the bit allocation, none of the channel data capacity must be "wasted" in sending specific bit allocation instructions. The bit allocation can have time or frequency resolution up to that of the information used to generate the allocation. This allows backward adaptive systems to simultaneously be more efficient in their use of the channel capacity and to have finer time or frequency resolution when required.

But the fact that the decoder must calculate the bit allocation entirely from information contained in the coefficient bit stream leads to the disadvantages of the backward adaptive bit allocation method. The information sent to the decoder has limited accuracy and therefore may contain small errors. In addition, since the decoder is intended to be low in cost, the calculation must be relatively simple, or it would drive up the cost of the decoder. Furthermore, the bit allocation algorithm becomes fixed as soon as the first decoders are deployed, and it then becomes impossible to update the psychoacoustic model.

Advantages claimed for the hybrid approach are that the modification data sent to the core bit allocation routine is substantially less than would be required for normal forward adaptation and that the psychoacoustic model can be updated dynamically. At the same time, the frequency resolution and time resolution can be as good as that of the analysis and synthesis filter banks.

MASKING MODEL

As we discussed in detail last time, the human auditory system is characterized by a masking model in which higher level tones suppress the audibility of lower level sounds that are nearby in frequency. The masking effect is more pronounced for sounds at frequencies higher than the higher level tones than it is at lower frequencies. It is also more pronounced for sounds closer in frequency to the higher level tones than it is for those further away. There are a number of critical bands in which the masking takes place, and the bands are wider in bandwidth at the higher frequencies.

In the AC-3 masking model, certain simplifications are made. These allow a masking curve such as shown in Figure 14.4 to be used. Here the downwards masking falls off very steeply with frequency, and there are two curves for upwards masking. One of these has a fast response to changes in signal level and falls off moderately quickly with respect to upward frequency differences. The other has a slower response, occurs at a somewhat lower level, but extends quite a bit higher with regard to frequency differences.

In the actual implementation of the AC-3 system, the downwards masking function is not realized in order to further simplify the system. This results in occasional inefficiency in coding but allows a significant reduction in complexity of the decoder.

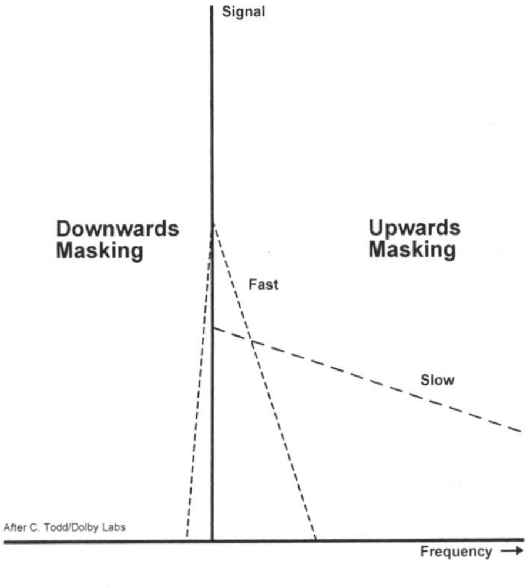

Figure 14.4 – Prototype Masking Curve

Table 14.1 – Dolby AC-3 Audio Coding System

Characteristic	Sampling Frequency		
	32 kHz	44.1 kHz	48 kHz
Audio Bandwidth	15 kHz	20.6 kHz	22.5 kHz
Normal Blocks			
Block Length	512 samples	512 samples	512 samples
Block Duration	16 ms	11.61 ms	10.66 ms
Number of Sub-bands	256	256	256
Sub-band Bandwidth (Fs/512)	62.5 Hz	86.133 Hz	93.75 Hz
Short Blocks			
Block Length	256 samples	256 samples	256 samples
Block Duration	8 ms	5.805 ms	5.333 ms
Number of Sub-bands	128	128	128
Sub-Band Bandwidth (Fs/256)	125 Hz	172.26 Hz	187.5 Hz
Block Repetition Rate	125 Hz	172.3 Hz	187.5 Hz
Block Repetition Period	8 ms	5.805 ms	5.333 ms
Bit Rate – 2 channel stereo	192 kb/s	192 kb/s	192 kb/s
Bit Rate – 5.1 channel surround	384 kb/s	384 kb/s	384 kb/s
Coding Modes	3/2, 3/1, 3/0, 2/2, 2/1, 2/0, 1/0, 1+1 tagging for additional multi-lingual or commentary channels & associated services		

AC-3 DETAILS

A summary of the characteristics of the Dolby AC-3 system is shown in Table 14.1. To the extent practical, this table follows the format of Table 13.1 for MPEG audio in order to make comparisons possible.

In the AC-3 system, frames consist of six blocks. The blocks are variable in length to accommodate different applications. In addition, the blocks are overlapped by 50 per cent on each side, resulting in each input sample being contained in exactly two blocks. Thus the first N/2 samples of each block are the identical samples contained in the last N/2 samples of the previous block. For the typical distribution application, blocks of 512 samples are used, equaling 5.333 average milliseconds per block after subtraction of the block overlap. Time resolution as low as 2.67 msec is possible.

The buffered samples are multiplied by a window function to reduce block boundary discontinuity effects on the spectral estimate provided by the later transform (filter) process. The window also improves the frequency analysis properties of the encoder.

Sub-band filtering is accomplished with a transform having 256 frequency domain points, resulting in 93.75 Hz spacing of the sub-bands. Except in a portion of the core bit allocation routine, all of the individual filters are used and the full frequency resolution of the filter bank is maintained. In the core bit allocation process, sub-bands are combined on a non-uniform basis that is intended to approximate the critical bands of the human auditory system.

The AC-3 encoder performs a time-to-frequency domain transformation using an oddly-stacked Time-Division Aliasing Cancellation (TDAC) technique which consists of a Modified Discrete Cosine Transform (MDCT). An advantage claimed for this approach is that the 50 per cent block overlap is achieved without requiring an increase in the resulting bit rate.

TDAC is a critically-sampled analysis technique, meaning that it generates exactly N unique non-zero transform coefficients on the average in an interval of time representing N input PCM samples. In TDAC, each MDCT transform of block size N generates only N/2 unique non-zero transform coefficients, so critical sampling is achieved with the 50 per cent block overlap already described. Conventional transforms (such as the DFT or standard DCT) preclude overlapped operation because each N-point transform generates N unique non-zero transform coefficients.

It is also claimed for the MDCT transform that several memory and computationally efficient techniques are available for its implementation, making it more cost effective than other methods. Bits are allocated to each sub-band based upon a comparison with the log-spectral envelope computed for each audio block, calculated on a critical-band frequency scale.

ENCODING OF THE SPECTRAL ENVELOPE

One of the processes that helps improve the transmission efficiency of the AC-3 system is the manner in which the spectral envelope data is encoded. The sub-band filters in the transform filter bank do not have perfect brick walls but instead fall off at a rather gentle 12 dB per adjacent filter (every 93.75 Hz). This results in the level differences between adjacent sub-bands rarely exceeding 12 dB.

By establishing a unit value of 6 dB per step, it is possible to differentially encode the spectral envelope from the lowest frequency sub-band through the highest, using only five possible values for each difference. The first coefficient's exponent, representing the sub-band beginning at zero frequency (hence the DC term), is sent as an absolute value. Then each succeeding, higher sub-band exponent is sent as a difference from the one below it. The only values required are +2, +1, 0, -1, and -2, representing +12 dB, +6 dB, 0 dB, -6 dB, and -12 dB change from the next lower sub-band.

In the simplest implementation, differentials are coded into 7-bit words each carrying the values for three sub-bands. In this instance, each exponent is coded in about 2 1/3 bits. The method is called D15, meaning that sub-bands are coded individually with five levels to represent each. D15 coding results in a very accurate spectral envelope.

Sending a D15 coded spectral envelope for every block would use too much of the data capacity, so D15 coding is only used when the spectral envelope doesn't change for a large number of blocks. In that case, the spectral envelope is only sent every 6 blocks or so (once per frame – about every 32 msec), resulting in a data rate for the spectral envelope of <0.39 bits per exponent. An example of the spectral envelope, coded over six audio frames, is shown in Figure 14.5. In this drawing, provided by Craig Todd of Dolby Laboratories, the fine lines are the individual spectra of the six frames, and the dark line is the coded spectral envelope.

When the spectrum of the signal is changing, the spectral envelope must be sent more frequently. To keep the amount of data that represents the spectral envelope from becoming too great, it can be coded with lower frequency resolution. There are two ways to do this. One, the D25 method, provides medium frequency resolution by sending the differential for every other

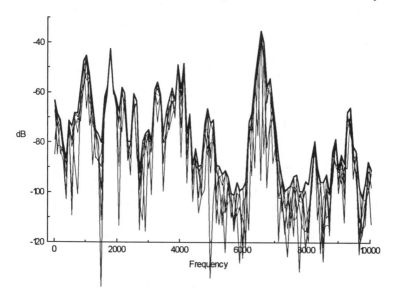

Figure 14.5 – Spectral Envelope, D15 Method
(Courtesy of Craig Todd/Dolby Laboratories)

frequency coefficient. This yields half the number of bits of the D15 scheme (1.16 bits/exponent) and half the frequency resolution. D25 coding is generally used when the spectrum holds for 2-to-3 audio frames before changing substantially.

The coarsest coding method is D45, in which difference values are sent for groups of 4 frequency coefficients. This results in a quarter the number of bits of the D15 method (0.58 bit/exponent) and, of course, one-quarter the frequency resolution. With the lower data volume per spectral envelope, D45 codes can be sent every audio block, offering finer time resolution for transient conditions.

Through selection of the exponent coding method, it is possible to change the balance between fine frequency resolution combined with coarse time resolution and coarse frequency resolution combined with fine time resolution as required by the material being coded. Associated with each input audio block, once it is coded, is a 2-bit value called exponent strategy. It has values representing D15, D25, D45, and REUSE.

Input audio frames, when coded, are carried in output data blocks that are grouped six at a time into data frames. Most of the time, a D15 strategy is used for the first data block in a data frame, and the remaining five blocks reuse the same spectral envelope information. When there are signal transients, the spectral envelope data is sent more often using one of the other strategies.

ADDITIONAL TECHNIQUES

Sometimes, when the system is operated at very low bit rates, signal conditions are possible in which the encoder would run out of bits. Under these circumstances, advantage is taken of the stereophonic irrelevance of the human auditory system that we discussed last time. The higher frequency channels, above a specific "coupling frequency," are coupled together and combined into a common "coupling channel." The coupling channel is treated similarly to the individual channels, including the formation and transmission of the spectral envelope and the quantized mantissas.

Below the coupling frequency, the channels are coded independently. Above the coupling frequency, "coupling coordinates" are calculated and transmitted for each channel. The coupling coordinates allow the level in each channel that was combined into the coupling channel to be adjusted in level relative to the coupling channel amplitude in order to closely approximate the original level in that channel.

There is a mixdown capability included in the AC-3 design that allows the 5 channels to be combined into two channels for stereo presentation or for matrix surround decoding and into one channel for monophonic presentation. The mixdown capability requires a small amount of additional processing in the decoder/receiver that has caused some disgruntlement on the part of certain receiver manufacturers. But it does allow the reproduction of all the sound information from all the channels without requiring special care in the production process to, for instance, avoid losing information from the surround channels when the program is reproduced in a mono output.

The AC-3 system deals with the issue of loudness differences between program segments. It does this by allowing input of the set point level of normal spoken dialogue in the center, dialogue channel and by coding that value into the data stream. This then serves as a dialogue level control for the decoder that allows it to adjust the playback volume so that the level of reproduced dialogue will be relatively constant for all programs and all channels.

Included with the other data for each coded audio data block (representing 5.3 msec of audio) there is a dynamic range control value. This can be set during the encoding process or at any time or place downstream to allow effective dynamic level control of the encoded material. The decoder uses the dynamic range control words to adjust the level of the reproduced signals up or down with roughly ¼-dB accuracy, applying a smoothing process to avoid the artifacts of stepped gain changes.

A compression algorithm can be used at any point in the system to create the dynamic level control data. Interestingly, the signal can be re-compressed, based on the original signal data, merely by replacing these control words. The default is for the decoder to use the dynamic level control data. But it is possible to moderate the effect in either the gain reduction or gain expansion directions or to ignore the dynamic level control data altogether. This partially or fully restores the original program dynamics for those listeners for whom this is appropriate.

CHOOSING A SYSTEM

In some situations, it may be necessary for users to select between the MPEG audio and Dolby AC-3 systems, either standing on their own or as part of overall system offerings by manufacturers. If you find that you need to make such a selection, be aware that the impact of the two techniques on perceived sound is quite subjective and depends heavily on the material being coded. Objective measurements will tell little that can be used to differentiate the systems. Should it become necessary, statistical subjective assessment is required, using listeners with good listening skills, if the superior of the two systems is to be identified. It should also be noted that the differences likely will only be apparent on high quality audio presentation systems.

PART 5

The Systems Layer

While digital video compression and digital audio compression are the prerequisite enablers of Advanced Television, it is the systems layer, especially as introduced by MPEG-2, that provides much of the functionality that makes ATV so attractive. In many ways, it is the packetization of the data to be transported that exists in the systems layer that is the real enabler for the widely discussed convergence of television, telecommunications, and computers.

Among the things that the systems layer supports is the multiplexing that permits combining multiple programs into a single channel. The first article in this part, Chapter 15, considers the systems that are necessary to carry out the multiplexing function both in a straightforward manner and in a more complex, and perhaps more efficient, manner called statistical multiplexing. It also begins to look at the issues surrounding the handling of multiplexed programs,

125

particularly with repect to the downstream insertion of material such as commercials.

The remaining three chapters in this part originally appeared as a three-part mini-series devoted to the MPEG-2 systems layer. They give a fairly full treatment to the packetization process, the kinds of streams, and the inner workings of the stream type that will see the most use in the electronic media – the transport stream. The use of tables to find one's way from the system multiplex to the program to the individual elementary stream is covered here.

These chapters also go further into the methods for switching MPEG-2 transport streams. They throw in the mapping of MPEG-2 packets into ATM cells for good measure. At the end is some personal speculation about the possibilities of carrying both compressed and raster-based signals in the same distribution systems by adopting packetization for the non-compressed signals.

By the time you finish this section, the vernacular of the systems layer (PESs, PIDs, adaptation fields, program map tables, and the like) should be familiar terminology.

15

Multicasting – Transmitting Multiple Programs per Channel

We have noted on several occasions in this book that our discussion covered applications ranging from HDTV to multicasting. But we have not yet explicitly looked at the technologies of and issues surrounding multicasting. Because of the importance of multicasting to cable, DBS, wireless cable, and possibly even broadcasting, in this chapter we will change that situation. We will start by defining multicasting and looking at the technologies that make multicasting possible. Then we can delve into some of the issues that must be addressed by system developers before multicasting can be truly practical to implement.

MULTICASTING DEFINED

Multicasting is a technique that permits transmitting several *program streams* in a single *channel*. For cable, wireless cable, or broadcasting, that channel will be 6 MHz of spectrum. For DBS and program distribution applications, the channel will be a transponder with a particular bandwidth. For wired connections on coaxial cable or fiber optics or for recording, it will be a designated amount of spectrum or a particular data capacity. It is important at the outset to differentiate the data that carries the program material, i.e. the program stream, from the portion of spectrum that will contain the modulated signal that will carry the data, i.e. the channel. Multicasting is made possible by the use of digital video compression (DVC).

First published September, 1993. Number 13 in the series.

- ## Compression/Transmission/Storage

- ## Reception/Recovery/Decompression

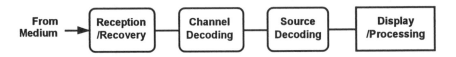

Figure 15.1 – Generic Model of a Digital Compression System

Figure 15.1 is the model we have been using consistently to describe digital video and audio systems in a generic way. Basically, the *source* is the studio system; *source coding* is the digital video or audio compression process; *channel coding* applies error correction, does various types of data formatting, and modulates the signal in a manner appropriate for the particular *medium*; and the *transmission* or *recording* process applies the channel coded signal to the medium, as, for instance, in a transmitter and antenna or a recording amplifier and recording head. The recovery side of the system applies the inverse of each of these functions in the reverse order. (For a more detailed description of each of the blocks, see the first article in this series, in Chapter 1.)

In a multicasting application, a number of sources are compressed in individual source coders (compressors), combined in a *multiplexer,* applied to a single channel coder, and sent through a single transmitter or record amp, into a single channel in the medium. This is shown in Figure 15.2. In general, on the receiving/recovery side of the system, the data is inverse channel coded, then demultiplexed. Usually a single program stream is then selected and sent to a decompressor (*inverse source coder*) where it is converted back into its source form for further processing or display. This is shown in the bottom half of Figure 15.2.

DATA CAPACITIES AND REQUIREMENTS

The channel to be used has a certain data capacity that is determined by the bandwidth, the form of modulation, the amount of error correction overhead applied, the overhead required for addressing and control, and similar system characteristics. Depending upon the system choices made, a 6 MHz television channel might provide 16-25 megabits per second (Mb/s) for terrestrial broadcasting, 21-43 Mb/s for cable, and 20-40 Mb/s for wireless cable. A 40 MHz satellite transponder will likely provide on the order of 40 Mb/s. The job of the multiplexer is to divide this

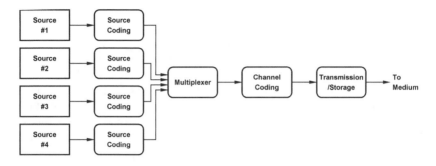

Compression/Multiplexing/Transmission/Storage

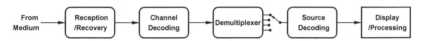

Reception/Recovery/Demultiplexing/Decompression

Figure 15.2 – Elements of a Generic Digital Multicasting System

data capacity and to allocate it to the several program streams that will be combined into the single channel.

In examining how the data capacity will be divided between sources, it is important to recognize that different types of sources have different requirements in order to achieve the quality desired for their applications. The more complex the image and the more motion the image contains, the higher the data rate that will be needed to yield results of a particular quality. As the technology improves, the data required for each type of program source is likely to decrease, allowing more programs to be carried within a given channel data capacity.

For purposes of our discussion, we will consider three types of program sources, and we will assign them values for the bit rates to adequately carry each. The scanning structure in each case is assumed to be 525 lines at 60 fields per second with component operation used. (See Chapter 32 for the reasons we avoid use of composite, or NTSC, operation.) The first type of program source uses full-motion video with lots of detail and lots of action – a basketball or hockey game, for example. The second is a film source, such as a movie, that starts at 24 frames per second and is subsequently scanned into video form. The third is a video source with limited motion and detail – for example, a lecturer at a blackboard.

To properly render the three sources with current compression technology, it might take approximately 5 Mb/s for the full-motion video source, 2.5 Mb/s for the film source, and 1.25 Mb/s for the limited video source, depending upon the quality to be achieved. The hierarchical values will make the following analysis easier, in any event, and should be considered just examples. These values include compressed stereo audio for the full-motion video and film sources and compressed mono audio for the limited video source. The bit rates are exclusive of the channel coding overheads mentioned previously.

MULTIPLEXING STRATEGIES

There are two strategies for multiplexing several program streams together into a single channel. The simplest is to use a fixed allocation of bits to each program stream. Thus, if five full-motion video sources were to be combined into a single channel with a net payload capacity of 25 Mb/s, each source coded stream would be allocated 5 Mb/s. If there were three full-motion video sources and four film sources to be combined, they, too, could fit within 25 Mb/s ((3*5)+(4*2.5)). Similarly one full-motion video source, six film sources, and four limited video sources could also be combined into 25 Mb/s (5+(6*2.5)+(4*1.25)). Many other combinations of these fixed values are possible.

The other, more sophisticated strategy is to use a variable allocation of bits to the various program streams to be combined. This variable or "statistical" multiplexing takes account of the fact that the complexity and amount of motion in a television image or film is constantly varying with changes in the scene content. On a statistical basis, several program streams that are combined into a single channel will not all require their maximum data rate at the same time. Therefore it is possible to share the channel data capacity between them on a variable basis and thereby improve the efficiency of the system. This can permit the use of more program streams or can allow the delivery of higher image quality for each of the program streams.

FIXED MULTIPLEXING

With fixed multiplexing, each compressor that feeds a particular multiplexer is pre-assigned a data rate that it will produce on its output. The data rate is determined by setting the frequency of a clock that controls the readout from a buffer that, in turn, is used to compensate for the fact that variable length coding is part of the compression process. There is then feedback of a measure of buffer fullness to a normalization matrix that determines how many bits of quantization will be used to represent the image at any instant. The greater the number of bits, the higher the fidelity of the reproduced image after decompression. The quantization level is controlled by the feedback in such a way that the average bit rate going into the buffer matches the constant bit rate being clocked out of the buffer. The result is that, as the complexity and motion in the compressed image increase, the fidelity of the reproduced image decreases. (This all was discussed in more detail in Chapter 12.) See Figure 15.3 for an example of part of a system based on fixed multiplexing.

The use of fixed multiplexing does not mean that the data rates of the various program streams cannot be changed. Rather, it means that the rates are only changed periodically, as, for instance, at the times of program changes. Thus assignments can be made to the various compressors feeding a particular multiplexer at the time the inputs to those compressors are changed from one

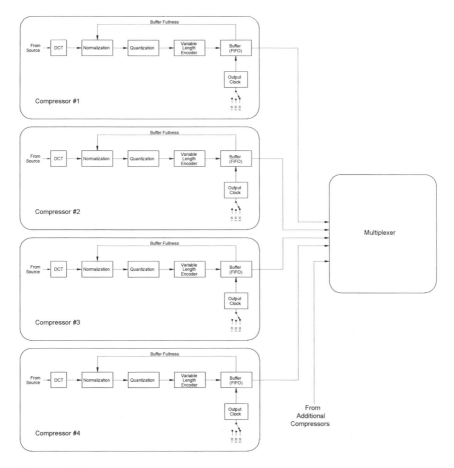

Figure 15.3 – Fixed Multiplexing, Simplified Example

program to another. The total of the data rates assigned cannot exceed the total data capacity of the channel.

It is helpful in understanding both fixed and statistical multiplexing to note that only the output clock frequency needs to be changed. The changes in normalization matrix then happen automatically as a result of the buffer fullness feedback.

STATISTICAL MULTIPLEXING

In statistical multiplexing, advantage is taken of the fact that it is unlikely that all the program streams being combined by a multiplexer will have maximum image requirements at the same time. Thus it becomes possible to dynamically allocate bits between compressors based on their contention with one another for the bit rates to satisfy the demands of their particular programs. This is done by feeding a measure of image complexity and motion from each compressor to the multiplexer. See

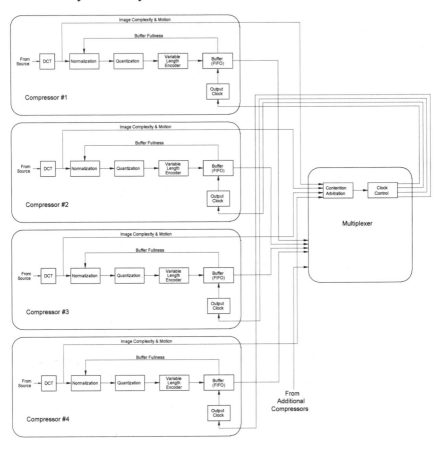

Figure 15.4 – Statistical Multiplexing, Simplified Example

Figure 15.4. The multiplexer then arbitrates between the connected compressors, controlling the data rate from each so that the total bit rate in the channel is divided between them based upon their relative needs at each instant. The data rate assignment is then dynamically reflected further back into each compressor through the buffer fullness feedback discussed previously.

With statistical multiplexing, overall efficiency of use of the channel capacity is improved. Program streams that can be carried at a high enough quality level by fewer bits give up some of their bits to program streams that require more at any moment. The quality of the more active and complex channels is thereby improved. The possibility for such quality improvements can be traded off to permit more program streams to be carried in the channel. Thus there is a balance to be achieved between the quality that is possible and the number of channels that are carried.

MULTIPLEXING ISSUES

The use of digital video compression combined with multiplexing allows the sharing of channel capacity between several program sources. This has the effect of multiplying the number of programs that can be sent in a given amount of spectrum, thereby multiplying its efficiency. Along with this efficiency, however, comes a significant level of complexity that is likely to have operational ramifications for many applications. In particular, those situations in which it is necessary to process the signals downstream – by adding commercials at a cable headend, for example – may require special techniques.

The requirements for cutting into a program stream to insert other material have been discussed in this column before. (See Chapter 12 for a more complete treatment.) If decoding and re-encoding of the program stream and the image quality degradations that go along with them are to be avoided, the material to be inserted must be compressed to match the data rate of the material to be cut into. The cuts are also likely to have to be made at black. This means that the material will have to be processed at the point of initial compression to include black holes for making later downstream insertions.

The use of different data rates for the different types of program streams that multiplexing permits only complicates this circumstance. The commercial, promo, or PSA to be inserted will certainly be stored in compressed form at the cable headend. Assuming fixed multiplexing for the moment, if it is desired to insert the same commercial into both video program streams at, say, 5 Mb/s and film program streams at, say, 2.5 Mb/s, does this mean that two copies of the commercial will have to be kept in the library at the headend? Probably. And will the information have to be provided to the commercial insertion system in advance to tell it which version to insert and to set up all of the data rates to do that properly? Certainly. Is there a way to avoid these complications? Maybe.

If a reserve is maintained in the overall data channel of, for instance, 2.5 Mb/s, all insertion material could be inserted at 5 Mb/s, so long as insertions were made into only one program stream at a time. If the program stream into which the insertion were to be made were already a 5 Mb/s stream, nothing would have to be done but the insertion. If it were a 2.5 Mb/s stream, the program stream would be increased to 5 Mb/s for the duration of the insertion. Then it would revert to 2.5 Mb/s after the insertion was completed. Of course, decoders would have to be able to follow these instantaneous data rate changes without a disturbance. It must also be recognized that the efficiency of channel utilization would go down by leaving the reserve for the insertions. It is assumed that lower data rate program streams would not have insertions made, or a larger reserve would have to be provided. Are there any better ways to do this? Hopefully.

What happens if statistical multiplexing is used? Now, there is no fixed rate of 5 Mb/s or 2.5 Mb/s to match because the data rates of individual program streams vary continuously as they contend for channel capacity with their channel-mates. If the commercial, promo, or PSA to be inserted were stored in uncompressed form – not really a practical thing to do, but instructive to consider – it might be possible to have a variable rate compressor that tracked the data rate of the program stream to be replaced by the insertion.

A more practical solution might be to force all program streams into which the insertions are to be made to a particular data rate some time prior to the point at which the insertions will be permitted. Then all downstream insertion material can be stored at that data rate and inserted as an exact replacement. This only works for statistical multiplexing, and it requires that the multiplexer be under control of the facility that does the post production to add the black holes, so that the multiplexer can be properly steered. It also requires that there be industry agreement on the data rate to be used for the insertions or else different pre-compressed versions of material for insertion will have to be maintained at headends to deal with signals from different networks.

The foregoing discussion only begins to explore some of the considerations that must go into the planning of systems using digital video compression and multiplexing. Careful analysis of all aspects of these techniques and the answering of many more questions are needed before the industry can begin to put systems based upon them into widespread use. Just as important will be industry standards for equipment, techniques, and operating practices that will enable maximum practical efficiency and minimum practical cost to be achieved.

16

Bits, Bytes, Packets, Headers, & Descriptors

The coming conversion to compressed digital video will bring with it a wholly new way of transporting signals from one piece of equipment to another and from one place to another. The sync pulses of analog video signals will be banished. Even the SAV (*S*tart of *A*ctive *V*ideo) and EAV (*E*nd of *A*ctive *V*ideo) synchronization words of current digital video techniques (based on CCIR Recommendation 601) may disappear. These will be replaced with mechanisms used in digital communications and computers.

The impact of this change will be widespread and profound. It will affect the way images and sound are stored and recovered, transmitted and received, distributed and switched. It will also enable the inclusion of television material in the more general flow of data and communications that will result from the eventual convergence of the computer, communications, and entertainment industries. The system to support this general flow of data and communications is what is popularly referred to as the Information Superhighway or in government parlance as the National Information Infrastructure.

This chapter we will look at the nature of the transportation systems that will be used to carry television signals in the new world of Advanced Television. We will work toward understanding the MPEG-2 and Grand Alliance "systems" or "transport" layers. In the next chapter, we will see how these map into the general communications scheme called Asynchronous

---

First published June, 1994. Number 19 in the series.

• Compression/Transmission/Storage

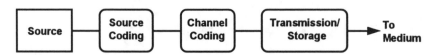

• Reception/Recovery/Decompression

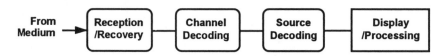

Figure 16.1 – Generic Model of a Digital Compression System

Transfer Mode (ATM), and we will look at the details of how the signals must be switched.

BITS, BYTES, AND PACKETS

Before examining how the new transportation systems will work, we should remind ourselves of how compression systems go together in a general sense, so that we can see where the transport piece fits into the overall picture. This is shown in Figure 16.1, the model we have used on many occasions to describe how compression systems work from a top level point of view.

The image (or sound) from the source is digitized and redundancy is removed (compression takes place) in the source coding block. The resulting data is then formatted for transmission, error correction coding (ECC) applied, and modulation performed in the channel coding block. Finally, the modulated signal is delivered to the storage or transmission medium for eventual delivery to the decoder where all of the processes are reversed. It is the formatting of the data for transmission at the beginning of the channel coding block that we are about to investigate.

The output from the source coder is a variety of different types of data that jointly comprise instructions to the decoder as to how to put the image or sound back together. Those different types of data include such things as motion vectors, discrete cosine transform coefficients, and quantization matrix selections or downloads for images, and power spectral density values and filter bank coefficients for sounds.

In order to get the various forms of data to the decoder in a manner that they can be identified and used, they must be grouped in such a way that they can be steered to the right place in the decoding process. They must also be able to be multiplexed with one another in such a way that they can share the same channel yet be separated efficiently at the receiving end of the channel.

Moreover, the decoder must be supplied with sufficient information to recognize what kind of data it is about to receive without having to fully decode the data in order to learn what it is. This is necessary so that the various elements of the decoder can spend their time efficiently processing the data that applies to them while ignoring data intended for other elements. This approach also permits the future addition to the data stream of new functions without interfering with the functions originally supported since those new functions can be ignored by the original decoder processes that were not set up to handle the new functions.

The method for accomplishing all this is to put the data in a defined order, in a group of data of a defined (although not necessarily fixed) size, and to attach to the front of that data codes that identify the associated data and provide synchronization for the decoding processes. The combination of these codes together with the data they identify and synchronize is called a **packet**. It is the use of packetization that is going to revolutionize data communications and support the convergence of the technologies and industries that is being so widely discussed.

PACKETS, HEADERS, DESCRIPTORS DEFINED

A packet can be defined as a grouping of serial data of a similar kind for transportation, storage, and eventual processing. A packet is introduced by a **header**, which can be defined as a preamble attached to the beginning of a packet to indentify the contents of the packet (the payload) and to provide services necessary to its use. Finally, a **descriptor** is an optional attachment to a header that provides additional information about the payload to permit its proper use.

A packet, then, provides a package in which similar data is grouped. For example, video data can be grouped in a packet that has a header identifying it as video. Audio data can be grouped in a packet that has a header identifying it as audio. The video data might also be grouped by the type of information it is; for instance motion compensated residual might be segregated into a packet with a particular header while motion vectors are put into another packet with a different header.

It is important to recognize that packets can be contained within packets. Since a packet is only a package, so long as the meaning of the header (the wrapper) is understood identically by all users, the contents can be defined in any manner desired. This allows packets to be assembled for a particular purpose, containing packets assembled for other purposes, and the two levels or "layers" of packets can work in complete harmony with one another.

It is the encapsulation of packets within packets that gives the packetization concept much of its power. It is what allows the carriage of MPEG-based programs over ATM networks in an efficient manner. It is what permits the intermixture of those MPEG programs with telephone calls and banking data. More importantly for the entertainment industry, it is one of the things that makes possible putting together high definition television (HDTV) and standard definition television (SDTV) signals in various combinations within a single transmission channel, thereby enabling so-called "flexible use," statistical multiplexing, 500 channel cable systems, 150 channel wireless cable systems, and the like.

PACKETS, PACKETS, PACKETS

Looking at television applications, packets separate and package the various forms of data (elementary data streams) such as video, audio, and associated or "private" data for sending them from their source encoders to their corresponding source decoders. Packets divide the elementary data streams from encoders to decoders into segments of a manageable size. Packets generically can be one of two types: variable length and fixed length.

Variable length packets tend to use the natural grouping of data from the coder. They tend to be relatively long. They generally are used for relatively error-free channels and media, for example, hard-wired networks and recording media such as CD-ROMs, tape, and the like.

Fixed length packets use an arbitrary length based on the needs of the transmission system through which they will be carried. They tend to be relatively short. They generally are used for noisy channels and media such as satellite transmission, terrestrial microwave, and terrestrial broadcast.

The packet headers provide identification of the contents of the packets along with necessary synchronization. Also included in the header may be an indicator of message length (for variable length packets) and other control functions (such as routing information in ATM systems).

Certainly the best defined and probably the most widely used of packetization systems for video and audio will be that provided by the MPEG-2 standards. MPEG (the Moving Pictures Expert Group of the International Standards Organization and the International Telecommunications Union) has had participation from hundreds of individuals, representing scores of companies, from dozens of countries around the world, all for the purpose of defining the techniques and the common language (syntax) to be used in communicating pictures and sound in compressed digital form. Most of the techniques and syntax used by the Grand Alliance for its HDTV system have been adopted wholesale from the MPEG standards. Consequently, we will use the MPEG-2 packet structure as our model in our further exploration of packets.

MPEG-2 PACKET STRUCTURE

The MPEG-2 packet structure exists at what is termed the "systems layer" of the MPEG-2 scheme. It consists of a number of different types of data streams that can be assembled to meet the needs of various applications.

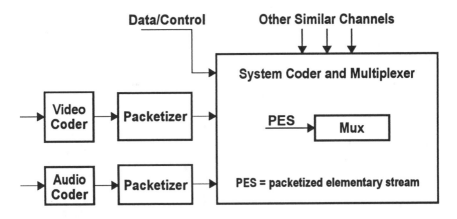

Figure 16.2 – Formation & Multiplexing of Packetized Elementary Streams

At the most basic level of the MPEG-2 systems layer are Packetized Elementary Streams (PESs) that carry one source of data for one application. Thus a PES could carry a video stream or one of several audio streams. The PESs are composed of the natural data segments from the related encoders plus headers added to identify the data in those segments to the decoders. The formation of the PESs takes place in packetizers associated with each encoder as shown in Figure 16.2.

PESs might be used to directly connect an encoder and a decoder, but more often than not they will form the basis for more complex data streams that will carry a multiplex of several PESs. There are two types of these more complex streams: Program Streams and Transport Streams. The relationships between Packetized Elementary Streams, Program Streams, and Transport Streams are shown in Figure 16.3. [This drawing is significantly modified from the original published with the article. With any luck, it's an improvement.] The Program Streams and Transport Streams are formed in specialized packetizers and multiplexers from PES inputs as shown in Figure 16.4.

Program Streams comprise one or a combination of PESs. A Program Stream permits joint use of multiple video and elementary streams. There is a common timebase among the PESs in a Program Stream so that, for example, a video PES and several audio PESs that are part of a single Program Stream can be decoded in lip sync with one another. Program Streams use variable length packets that contain the naturally grouped data from the elementary streams. Therefore they are used for essentially error-free channels and for software processing.

It is possible to have multiple Program Streams that share common elementary streams without the necessity of sending the PESs multiple times. This is because Program Streams really consist of an addressing mechanism for

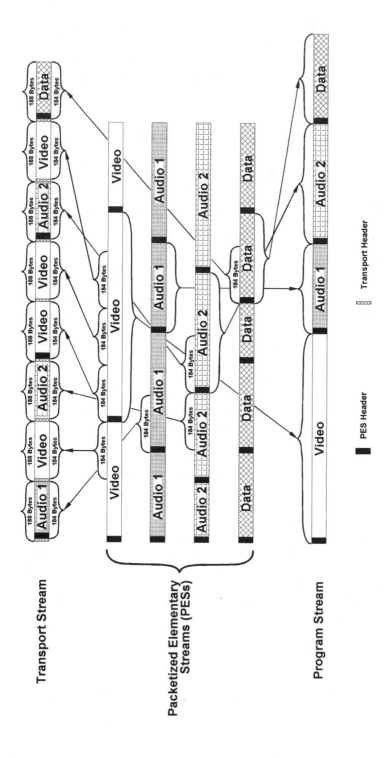

Figure 16.3 – Derivation of Program Streams and Transport Streams from PESs

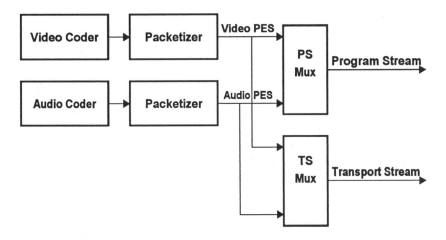

Figure 16.4 – Formation of Program Streams & Transport Streams from PESs

selecting among multiplexed PES packets rather than being physical entities requiring their own copies of the PESs. Thus it is possible to have Program Streams that share a video signal and have separate stereo audio programs for different languages. The video packets would appear once in the multiplex interspersed with all of the audio packets for the different languages. Similarly, it is possible to include multiple video signals and one or several related audio signals if the application is to offer alternate views of a scene with a common audio program. The several Program Streams would assemble the appropriate grouping of PESs from the overall multiplex through the addressing system (which we will discuss in more detail later).

Transport Streams, on the other hand, have many of the same characteristics as Program Streams, but they use fixed length packets derived from the same PES packets that would be used to form Program Streams. Transport Streams also comprise one or a combination of PESs that can be shared with other Transport Streams through an addressing mechanism and without the need for their repetition. There is also a common timebase for PESs in a Transport Stream. And the examples given above for reuse of PESs in Program Streams apply equally well to Transport Streams.

Because of their fixed length packets, Transport Streams are suited to noisy channels where robust communications is required. This is because it is relatively easy to add error correction coding to relatively small, fixed length segments of data and much harder to do so for longer, variable length data segments.

MPEG-2 Transport Stream packets total 188 bytes long. They divide PES packets into segments of 184 bytes and add a 4 byte header to each segment. There is also provision for an optional, variable length Adaptation Header to provide additional information to help in use of the data in the current and following packets in the particular PES within which they are contained. Adaptation Headers always follow a Transport Stream header and subtract from the 184 byte space available for the PES payload in the packet in which they are carried. They may also extend over more than one packet.

Following the header and any adaptation header, the payload of a Transport Stream packet consists simply of the next group of bytes from the PES. Periodically, this includes the PES header. The PES header always follows a Transport Stream header (and any Adaptation Header that shares the packet). If the previous packet carrying that PES was not filled by the data just before the PES header, it is bit-stuffed so that the PES header lines up with the beginning of the next Transport Stream packet payload.

Next time, we will start by looking at the Transport Stream packet header in more detail. We will then examine the addressing mechanisms provided by the Program IDs, the Program Association Table, and the Program Map Table. We will also look at the implications of all of this for switching and cutting the compressed signals, for fitting into the wider communications environment, and for the possible future use of a packet structure for raster-based images.

17

Header Details, PIDs, Adaptation Fields, & Tables

In our last episode, we met the principal characters in our tale of bits, bytes, packets, headers, and descriptors. We saw that a **packet** can be defined as a grouping (package) of serial data of a similar kind for transportation, storage, and eventual processing. A packet is introduced by a **header**, which can be defined as a preamble (wrapper) attached to the beginning of a packet to indentify the contents of the packet (the payload) and to provide services necessary to its use. Finally, a **descriptor** is an optional attachment to a data stream that provides additional information about the payload to permit its proper use.

We also introduced the MPEG-2 systems layer, the part of the MPEG-2 suite of standards that describes packetization and transport of the data. We decided to use MPEG-2 as a model for looking at packetization because it is so well defined and is likely to see very widespread use. The MPEG-2 systems layer supports two different methods for connecting encoders to decoders, namely the Program Stream and the Transport Stream, intended for different types of applications. We saw that both create packets through use of Packetized Elementary Streams (PESs) of video, audio, and private data, but that the structures of the resulting packets and the ways they are multiplexed are different. The relationships between the several types of packets and streams were shown in Figure 16.3 in the last chapter.

First published July, 1994. Number 20 in the series.

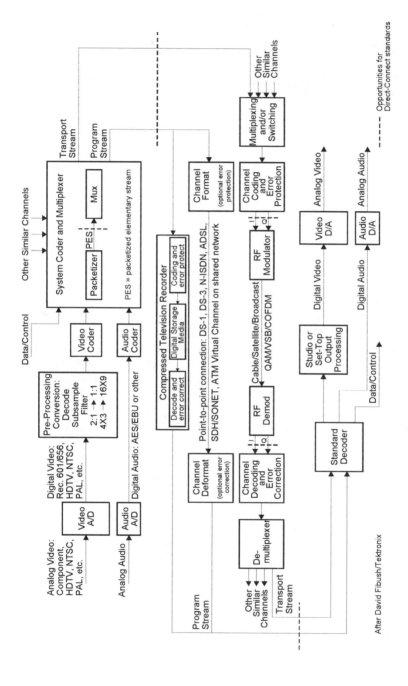

Figure 17.1 – Model of Generic Television System Showing Application of MPEG-2 Program Streams & Transport Streams

MPEG-2 SYSTEMS LAYER

We described, last time, the different reasons for using Program Streams and Transport Streams. A model that clarifies the applications for these packetization and transportation methods is shown in Figure 17.1. This drawing of a generic television system was conceived by David Fibush of Tektronix, enhanced by your intrepid writer, and is currently serving as the basis for discussions within a number of industry committees (especially SMPTE) concerning the places where standardized interfaces will be required and the characteristics of those interfaces. In the drawing, candidate locations for standardized interfaces are shown as dashed lines cutting across interconnections between boxes.

The Program Stream method uses the relatively long, variable length packets typical of elementary streams. Program Streams are appropriate for relatively error-free transmission channels or storage media. The Transport Stream method is used to provide robust communication of the data in noisy or distorted channels through the use, later in the channel coding process, of powerful error correction techniques that are more efficiently applied to its fixed length, relatively short packets. Because they embody the comprehensive range of MPEG-2 systems layer techniques, some of which do not apply to Program Streams, we will concentrate on Transport Streams for the remainder of our discussion.

TRANSPORT STREAMS AND HEADERS

The MPEG-2 Transport Stream includes packets carrying PES data and packets carrying directory tables that identify the current content of the signal multiplex and that associate PESs into programs. Critical to sorting all of this out is the 4-byte header that is positioned at the beginning of each Transport Stream packet, the remainder consisting of 184 bytes of payload data for a total of 188 bytes per packet. The four bytes of the header carry eight different functions that serve to identify the packet's contents and aid in its utilization.

The eight functions of a Transport Stream packet header (with the number of bits assigned to each function in parentheses) are: a sync byte (8), a transport error indicator (1), a PES packet-start indicator (1), a transport priority indicator (1), program identification – PID (13), transport scrambling control indicator (2), adaptation field control (2), and continuity counter (4). Of these, the PID is the one most important to understand in figuring out how Transport Streams really work. The layout of the Transport Stream packet header is shown in Figure 17.2.

PROGRAM IDENTIFICATION – PID

Since the PID has 13 bits assigned to it, it can take any of 8,192 (2^{13}) different values. Each PID value can be associated with a different packetized elementary stream (PES). (Think of the PID as an address for a PES.)

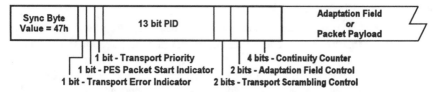

Figure 17.2 – MPEG-2 Transport Stream Packet Header Format

Remember that each PES represents one particular elementary stream — for example, a compressed video stream, a compressed audio channel, or some closed captioning data. There are a few PID values that are reserved for specific functions or for future definition. Thus, at any given moment, there can be up to about 8,175 different PESs in a single MPEG-2 Transport Stream. This gives the MPEG-2 Transport Stream multiplex ample address space to carry a great many different channels of material.

PROGRAM ASSOCIATION & MAP TABLES

In the MPEG-2 standard, there is no fixed relationship between any particular PID value and a given type of PES; PIDs are assigned to PESs on a temporary basis. Consequently, it is necessary to have an indexing system that dynamically defines the content of the Transport Stream. This is done through a pair of tables that are themselves carried within packets identified by PIDs.[12]

The first and highest level table, the Program Association Table, is always carried in packets with a PID value of zero (PID 0). The Program Association Table carries a list of program numbers for programs that are currently contained in the Transport Stream multiplex, together with pointers to other PID values that carry Program Map Tables for the respective programs. Program number 0 is reserved to carry a Network PID that points to the location (PID value) of a Network Information Table that, in turn, defines the meaning of all the other program numbers.

The Program Map Tables provide the mapping between program numbers and the elementary streams that comprise them. Such a mapping for an individual program is called a **program definition**. The program definition carries the PID numbers for each of the PESs associated with its program. It is this structure that allows multiple programs to share common PESs. For instance, two programs, each having its own program definition contained in a separate Program Map Table with an associated PID number, could both use a

---

[12] Supplemental implementation mechanisms that establish certain categorical relationships between PES types and PID values have been documented by the Advanced Television Systems Committee (ATSC) and by the European Digital Video Broadcasting (DVB) group. The several tables described are nonetheless required to make the system work.

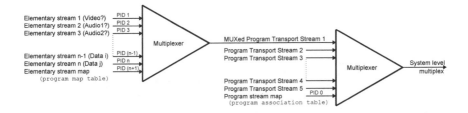

Figure 17.3 – Functional Model of MPEG-2 Transport Stream Multiplexing

common video PES while having separate stereo audio PESs in different languages. Both Program Map Tables would include the one video PES's PID in their lists along with the PIDs for their individual audio PESs.

The multiplexing of the elementary streams, PIDs, Program Map Tables, and Program Association Table is shown functionally in Figure 17.3, which is modelled after a pair of diagrams included in the Grand Alliance HDTV System Specification but applies to all MPEG-2 systems. Note that separate multiplexers are not required (but may be used) in real systems; they do serve to illustrate the relationships, however.

MULTIPLEXING AND DEMULTIPLEXING

The structure described also allows combining multiple preassembled programs with their own Transport Streams into a single new Transport Stream in a channel having sufficient bandwidth. To do so, the contents of the two (or more) Program Association Tables are combined into a new table at location PID 0. But care is required in doing this. Since every PES in a multiplex requires a unique PID, either distinct PIDs must have been assigned to the PESs when they were preassembled, or some of them must be changed on the fly as the new multiplex is assembled. Since preassignment of unique PIDs will probably be difficult to maintain, especially when the compressed material is prerecorded, it is likely that the PIDs will have to be altered as the new multiplex is put together. This is not particularly difficult to accomplish but must be considered in system designs.

With this structure, at a decoder, there is a three-step process to recover a particular program. First, the decoder must get the Program Association Table from PID 0 and look up the program number it wishes to decode. Then it must get the Program Map Table from the PID identified for that program in the Program Association Table. With the Program Map Table recovered, it can set the PES addresses (PID values) to be recovered by each of the stream decoders (video, audio, closed captioning, etc.). This effectively demultiplexes the overall Transport Stream, extracting the program desired. This process is shown functionally in Figure 17.4, again based on the Grand Alliance HDTV System Specification but applying to all MPEG-2 systems. Once again, real implementations are likely to be more efficient through the combining of

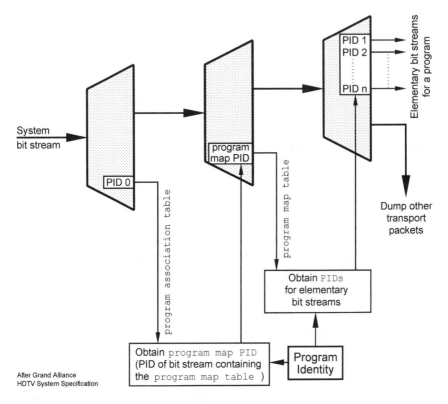

Figure 17.4 – Functional Model of MPEG-2 Transport Stream Demultiplexing

functions, but the drawing serves to show the minimum functionality required to recover any particular application bit stream.

ADAPTATION FIELDS

In addition to the standard header that appears at the start of every packet, some Transport Stream packets also carry an Adaptation Field immediately following the header. The presence of the Adaptation Field is indicated by the adaptation field control bits in the header. The Adaptation Field is a variable length data structure that carries additional control information to aid in the downstream processing of the packetized elementary stream (PES) in which it appears. At the beginning of an Adaptation Field is a fixed length component that includes a length indicator for the Adaptation Field plus a number of flags. The flags either directly carry information, or they indicate the presence of several optional additional fields in the variable length portion of the Adaptation Field that follows the fixed length portion.

Included in the Adaptation Field are several flags and a counter that assist in splicing pre-compressed program segments together plus some clock

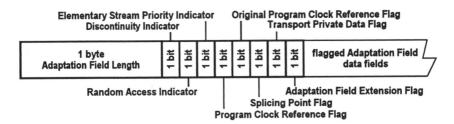

Figure 17.5 – MPEG-2 Adaptation Field Structure

Included in the Adaptation Field are several flags and a counter that assist in splicing pre-compressed program segments together plus some clock reference data that helps keep the decoding processes synchronized to the source. Among the flags provided are an indicator that a random access entry point into the PES bit stream is contained within that packet's payload data, an indicator that there is a discontinuity in one of the clock references from that packet onwards as would occur when bit streams are spliced, an indicator that a splicing point countdown field is present in the variable length portion, and indicators that two different types of clock reference fields are present in the variable length portion. The clock reference fields provide data about the phase of a 27 MHz clock that is used to synchronize the decoder to the encoder and will normally appear in only one of the PIDs associated with a particular program. The structure of the Adaptation Field is shown in Figure 17.5. The significance of these flags and fields is that they permit downstream equipment to perform splicing operations on a program multiplex without the necessity of decoding the PES packets.

SPLICING

Aside from supporting the flexible multiplexing of different PESs into programs and programs into overall system channels, probably the most important function of the systems layer is support of the splicing operations necessary to assemble program segments in succession and to insert replacement material downstream of that assembly. This cutting and switching can be thought of as being akin to assembly editing and to insert editing respectively. The Adaptation Field functions support these operations.

Because of the requirement that PES headers always occur immediately following a Transport Stream header (and Adaptation Field, if present), the required points for changes will always be on Transport Stream packet boundaries. The random access indicator flag indicates the packets that contain appropriate PES entry points. Since program segments that are to be joined must be properly phased to one another for a smooth transition and since this cannot be done instantly in practical equipment, a countdown function is provided to avoid the need for very long buffers to take up the slack while the

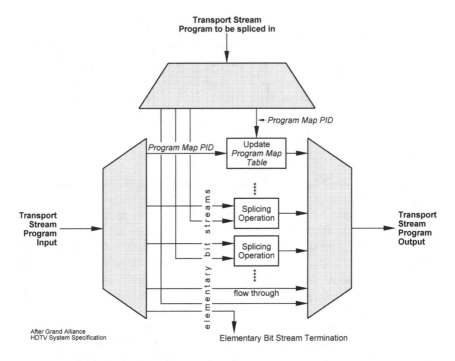

Figure 17.6 – Model of Functionality for Transport Stream Splicing

through zero in the packet preceding the one in which the new stream should be entered. The countdown itself is in a field in the variable length portion of the Adaptation Field whose presence is indicated by the splicing point flag.

When a splice is made from one program segment to another, an unexpected change will most probably occur in the clock reference information that is sent to the decoder in the Adaptation Field. This results from the derivation of the clock reference information from the separate encodings of each segment that were likely not timed to one another, especially for pre-encoded material. The discontinuity indicator flag is therefore set when a splice is made to indicate to the decoder that the discontinuity in the clock reference is not an error and that the decoder's clock phase should be updated. This will prevent a full relocking of the decoder that might otherwise occur.

The actual process of splicing program segments takes place in a system with the functionality shown in Figure 17.6. First the PES packets for the program to be switched must be extracted from the input Transport Stream by demultiplexing the PIDs identified by the Program Map Table. At the same time, some elementary streams can be passed through to the output unmodified while others can be dropped or added. Each of the elementary streams is spliced in a separate splicing operation that makes the packet substitution at an appropriate point for its input streams. As the various elementary stream splices occur, the Program Map Table must be updated to reflect the new data (probably including new PIDs) for each of the elementary streams now related to the particular program. At the time a splice occurs in the elementary stream carrying the clock reference information for the program, the discontinuity indicator flag must be set.

COMPLICATIONS

The method for splicing between segments just described leads to a number of complications that can impact either the complexity of the equipment required to make the edits in the program or the quality of the program produced at the output of the decoder. These complications become even more challenging when the splicing is done for the purpose of inserting segments (e.g. commercials) downstream of the assembly point, for example, at a network affiliate or cable headend. In such cases, the need to meet the timing of a program flowing through the switching location at both the entry and exit points of the program with a replacement segment provides opportunities for freezing the image, muting the sound, or even forcing a relocking of the entire decoding process.

How to avoid these hiccups is where we will begin in our next outing. We will also look at how MPEG-2 Transport Streams will fit into the cells of the ATM data communications systems that are likely to become prevalent in telecommunications and cable television distribution plants. We will examine the possibilities for using packetization for raster-based television images. And we will explore some possibilities for enhanced distribution signals to aid with downstream processing.

Making Packetized Television Work

We've come a long way in our exploration of the world of bits, bytes, packets, headers, and descriptors. In our last two installments, we defined packets and showed how headers are preambles to the data within packets; we saw how packets could be contained within other packets (wrapping one inside another); we looked at the structure of the MPEG-2 "Transport Stream" as one of the better examples of packetization and multiplexing of program material; and we briefly investigated what is involved in switching or splicing MPEG-2 packetized data.

This time, we will consider some of the practical implications of a number of these techniques and will explore how MPEG-2 is likely to fit into the larger world that will become the "information superhighway." We will also ponder how standard, raster-based television might ultimately fit together with compressed digital video. Just as a reminder, our discussion encompasses both Standard Definition Television (SDTV) and High Definition Television (HDTV) applications in all of the discourse to follow.

VARIABLE DATA RATES

As we've seen in some of our earlier exploration of the source coding of video, several techniques are used that cause the amount of data associated with different frames in a video sequence to vary. A prime example of this is the coding of some frames on an *intra-frame* basis ("I-frames"), some frames on a

---

First published August, 1994. Number 21 in the series.

predicted basis ("P-frames"), and some frames on a *bi-directionally* predicted basis ("B-frames").

I-frames require the largest amount of data to represent them because they are coded independently of any other frames in the sequence and must stand on their own. P-frames require substantially less data than I-frames because they result in only the differences between the prediction from the preceding I-frame or P-frame and the actual image being coded. B-frames require the least amount of data, since they are predicted from either the preceding or succeeding frame, with only the differences between the best predictions and the actual image being coded.

In addition to the different methods for processing the frames, variable length coding is also applied to the resulting data as a means to improve the efficiency of communications between encoder and decoder. It results in further variability in the amount of data associated with different frames in a video sequence.

In order to smooth out the actual data rate that is transmitted to something constant (or to a dynamically controlled value, in the case of statistical multiplexing), a first-in, first-out (FIFO) *buffer* is used. As we saw in an earlier article, a measure of the buffer fullness is used as a feedback mechanism to control the quantization level in the coding process so that the *average* data rate from the coding process matches the transmission data rate at the output of the buffer. We also noted that the quality of the coding increases and decreases depending upon the complexity and the amount of motion in the image at any particular time.

Underlying all of this is the fact that different coded data rates will be used for different Advanced Television applications. Thus for SDTV, something under 1.5 Megabits/second (Mb/s) might be used for "VHS quality" movies, and somewhere between 5 and 8 Mb/s might be the choice for "full motion video," while for HDTV, from 10 to 45 Mb/s are allowed for the Grand Alliance (GA) system.

As we saw previously, MPEG-2 Transport Stream packets have 188 bytes, of which 4 are the header and 184 are the payload. We should point out that the MPEG-2 standards allow for any coded data rate desired, within certain upper bounds for each profile and level (which are explained in Chapter 5). Since the number of packets per second is related to the coded data rate with some overhead thrown in, it will differ from system-to-system and application-to-application depending upon the choice made for the coded data rate.

ASSEMBLING SEGMENTS

So, what happens when we try to assemble program segments into a continuous program stream for delivery to viewers? Does the variability of the amount of data associated with individual frames affect how the joining of segments must be done? Are there any limitations on the kinds of effects that can be applied at transitions? The answers to these and related questions will, in large measure, determine the kinds of systems we will build for program integration in the future. Their answers all depend upon what it is that you want to do at the transition points.

There are two ways to handle transitions between program segments. One is to make the transitions in the raster-based domain — that is, using unencoded or decoded signals. The other is to make the transitions using fully or partially encoded signals. In the first instance, signals at the switching point are essentially the same as have been used throughout television history — they have pixels, scanning lines, fields, frames, and so on. They can be switched in the same manner as has always been the case, either in component or composite form, and then encoded following switching. This is what is being done at the huge new DirecTV uplink facility in Castle Rock, CO, for example. (Note that component storage and switching are to be preferred since the compression systems are component, and NTSC artifacts only add to the difficulty of encoding with high quality. But that's another subject.)

There are two principal problems in using unencoded or decoded signals. First, unencoded signals require vast amounts of storage capacity and wide data bandwidths for transportation. One of the main advantages of digital video compression is the economy it can bring to storage and transport. Second, decoded signals imply concatenation of encoding processes. Such decoding and reencoding has so far resulted in relatively low image quality in the final product. Later, we will discuss some possibilities for improving this situation.

For many installations, the only choice will be switching between pre-compressed program segments. Examples are likely to be video servers at cable headends or telco central offices, commercial insertion equipment at broadcast stations, cable and wireless cable headends, and centralized video libraries connected to the distribution channels through fiber networks. For these implementations, the techniques to permit direct switching of pre-compressed program segments will be required.

We saw in last chapter's installment that the Adaptation Header in MPEG-2 provides some facilities to aid in splicing program segments together. These included the splice countdown, the discontinuity indicator, and the random access indicator. These are all necessary to the process but probably not sufficient to do the whole job.

The problem is that program schedules are put together based on one second time increments, and compressed digital video does not march to that cadence. Neither, for that matter, does NTSC video. But with NTSC, we have the ability to switch on any field boundary (so long as the two signals are synchronized with one another), and field boundaries always come along within a sixtieth of a second (or within a thirtieth if we insist on always switching on the same field).

With compressed digital video, switching points can be far less frequent, and they may be highly irregular in their occurrence. While entry points for purposes of establishing lock for decoding synchronization or for recovering from an uncorrectable error occur at every I-frame, perhaps two or three times a second, an additional condition must be met for error-free switching. That condition is that the buffer must be filled to its mid-point (the target buffer fullness value) coincidentally with the start of an I-frame.

The reason for requiring the buffer to be filled to its mid-point is that the encoder has no way to know that a switch is going to occur downstream at any particular entry point. If a switch is allowed to happen with the buffer a little fuller than its target value and several such switches occur in succession, the buffers in all the decoders that are using the signal will eventually overflow, forcing the decoders to relock to the signal. Similarly, if several switches occur with the buffer a little less full than its target value, the buffers in all the decoders will eventually underflow (empty), again forcing the decoders to relock. Such a relock looks very similar to a channel change — not very desirable in the middle of a program.

It is important to note that the type of transitions we have been discussing are "cuts only." If anything with more production value (e.g., a dissolve, wipe, or key) is required, decoding of the signal will be necessary. The only thing that will be possible to help in making the switches look smoother with this type of switching will be to have fades-from-black and fades-to-black built into the segment starts and ends, respectively.

It should also be noted that nothing so far has forced the segments to occupy known amounts of time so that a schedule can be maintained. If the output of a video server is providing video-on-demand or near video-on-demand, this is probably fine. If some sort of program schedule is involved, this is probably a problem. We'll discuss a solution after we look at not just assembling program segments but inserting them.

INSERTING SEGMENTS

Inserting segments such as commercials or promotional announcements is similar to assembling segments, with the additional consideration that transition points must be matched at both ends of the inserted material with corresponding points in the program segment into which it is inserted. This matching must be achieved both for the video and the audio, which ideally will have entry points that are program-time coincident with those in the video. They may occur at different data stream times, however, and must be coordinated.

It is relatively straightforward to match the in-going point through the use of the splice countdown in the program's adaptation header and buffers in the switching hardware. But matching the out-going point depends on a number of factors related to how the two program segments were encoded.

It should be obvious that, if a program has a coded data rate of 1.5 Mb/s, we cannot simply insert a commercial with a coded data rate of 8 Mb/s. But what happens if we have a commercial with a coded data rate of 1.6 Mb/s or 1.4 Mb/s? There are two parts to the problem: (1) what happens in the overall data stream from encoder to decoder and (2) what happens to the buffers in the decoders.

The data stream result depends on whether there is unused capacity that can be allocated to the commercial, especially when it has a higher coded data rate than the program, and on the capability of the insertion multiplexer to handle dynamic data rate allocations through its packet switching functionality. The decoder buffers have the problems of overflow and underflow that we described in considering assembly of segments.

The underflow case can be handled through the use of "stuffing packets" to make up any shortfall when the coded data rate of the insertion is lower than that of the program. This will not contribute any improvement to the image quality, however, which will be lower than it might have been if the full coded data rate of the program itself had been used for the insert. When the coded data rate of the insertion is higher than that of the program, overflow will eventually result, leading to the associated picture disturbances.

The way to avoid any of these problems is for there to be strict control of the number of frames in a commercial or other insert of a given length, of the coded data rates used, and of the number of packets associated with the period of the insert. This number of packets must be supplied within the on-going program data stream and must be replaced with a like number of packets carrying the inserted material. None of these factors, however, is specified in either the MPEG-2 or the Grand Alliance system definitions.

PRODUCTION STANDARDS

All of which leads us to the need for production standards for compressed digital video systems. In all of the work that has gone into developing and defining MPEG-2 and the GA digital terrestrial television broadcasting system, no one has yet formally addressed the kinds of issues we've been discussing here. Before we can have practical implementations involving program integration downstream of the encoder, these issues must be resolved.

This says that fairly soon, the industry will have to grapple with adopting production standards covering the parameters to be used in the encoding and multiplexing processes. The Grand Alliance has identified many aspects of the matter in the Transport section of its "HDTV System Specification" document. Some of this has been in response to input from SMPTE committees seeking to make sure that things will eventually fit together.

SMPTE has recently launched an activity to look at the overall requirements for standards covering everything from interfaces between equipment to the kinds of parametric details we have been exploring. (The generic system model in Figure 17.1 in the last chapter originated as part of the thinking that led to this SMPTE effort.) The SMPTE activity will eventually turn into a series of efforts to define specific standards. Until this work is done and accepted by the many industry segments, implementation of compressed digital video can be expected to be rather chaotic.

ENHANCED DELIVERY

In our discussion to this point, we have assumed the desirability of switching in the compressed video/audio domain. This has been because of the combination of the simplicity offered by not having to decode and concerns about the artifacts introduced by decoding and re-encoding. The resulting limitation has been that we could do "cuts only" switching, and then only at specific, infrequent points in the program. There will be many instances, however, in which it will be desirable to do more than just cuts and to be able to place effects at arbitrary times within a program. Thus it will become necessary to decode and re-encode in these cases.

Can something be done to improve the resulting quality when decoding and re-encoding are required? Quite possibly, but not without an associated additional cost in the overall system. That cost will be the use of additional data bandwidth to deliver the programs to the points at which downstream program integration is required. The additional data bandwidth will be used to carry extra information to support exactly these functions.

For simplicity of the overall system, the ideal mechanism for carrying the extra information might be to transmit the normal, fully encoded signal along with some additional packets carrying "helper" information. In those installations where downstream processing was not required, the packets containing the fully compressed program could simply be separated and sent into the distribution system unaltered. The "helper" packets would then just be dropped out of the multiplex.

For those installations or times when downstream processing was needed, decoding could be done using a combination of the fully encoded signal and the helper information. The helper data might include quantization extensions for the DCT-transformed displaced frame difference data, more accurate motion vectors, or other enhancements that would allow higher quality decoding to be followed by subsequent re-encoding.

Use of the helper approach most likely would not be as efficient as creating a single signal with the same parameters and resulting quality. But the ability to easily separate a fully compressed signal for distribution to viewers without the need for an expensive encoder in smaller applications such as cable headends makes this an attractive method when considering overall system complexity and cost.

Enhanced delivery for downstream processing is something that has been discussed for quite some time within the Advanced Television community. But it is not included in MPEG-2, and it is not included in the Grand Alliance system. Any work going on in laboratories has not yet been announced. Yet these techniques may be required before implementation can really begin. Certainly there will have to be methods developed before downstream processing equipment can be marketed. These techniques may also have to be in place before commitments are made to the number of bits to be used for each program in the delivery systems to be built.

ATM SYSTEMS

Speaking of delivery systems, no discussion of packetized transport would be complete without recognizing the role likely to be played by ATM. ATM stands for "asynchronous transfer mode" and is rapidly becoming as universally accepted within the telecommunications community as MPEG-2 is within the video compression community. Efforts to develop ATM are just as widespread and energetic as those for MPEG, and it is likely to become the internationally accepted mechanism for high speed data communication.

ATM is a very large subject, and we will give it a full examination in future columns. The reason for mentioning it here is that an important part of recent discussions on the future information infrastructure has been concerned with the *interoperability* of MPEG-2 and the GA system with ATM. Many of the networks currently being planned for interconnection of video servers with cable headends or telco cental offices involve the use of ATM to carry MPEG-2 compressed digital video to the distribution points. Some schemes even propose to deliver MPEG-2 all the way to the home or office using ATM connections.

Suffice it to say that ATM is another packetization system that uses 53-byte packets (called "cells") having 5-byte headers and 48-byte payloads. It is intended for transportation of a wide range of data types including local and wide area network (LAN and WAN) data, telephone communications, compressed digital video ranging from video conference level to HDTV, and anything else that can be imagined.

ATM can interoperate with MPEG-2 and GA systems through the *nesting* (containing one packet within another) of packet structures we have discussed previously. This might be done by dividing the 188-byte MPEG-2 Transport Stream packets (including the 4-byte header) into four sections of 47 bytes each or by dividing the 184-byte MPEG-2 packet payloads into four sections of 46 bytes each. Each section will then fit within one ATM cell with one or two bytes left over for what is known as the ATM Adaptation Layer (AAL). Other schemes are also under consideration. The case with the MPEG-2 header included and one AAL byte is shown in Figure 18.1. We will look at the various methods and their implications when we consider ATM as a separate subject in the future.

UNIVERSAL HEADER

One of the benefits of the use of packetization is the ability to intermix many different types of data within a single data stream. Another is the ability to embed one type of packet within a different type, thereby allowing the interconnection of different systems as we've just seen with MPEG-2 and ATM. These generally support the interoperability sought for future systems. One of the problems with this type of relationship is that such systems are all designed assuming that their own transport systems are in control and that their

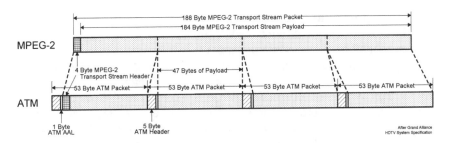

Figure 18.1 – MPEG-2 Transport Stream Packet Mapped
  Into Four ATM Packets

packet headers will always be used as the starting point for demultiplexing, but this is not always the case. This is called the "who's on top?" problem.

Another feature sought for future systems is *extensibility* — the ability to add capabilities over time without interfering with the functionality built into equipment that was produced before the new capabilities were defined. Extensibility is supported if packet-based equipment is designed to simply ignore data in the multiplex that carries header identifications that it does not recognize. Even better is the situation in which new header names can be recognized and the equipment updated to support the new functions identified.

We've seen a number of mechanisms in which different data types are given specific header values by systems. Now, suppose we have some data that we need to communicate through alternate or cascaded communications channels from time-to-time — perhaps time code, just to pick an example, but orders for commercial bookings would be just as appropriate. Let's say we need to send our time code first through a private data channel in an MPEG-2 Transport Stream, then through an ATM system, then through a third system. With our who's on top problem, each system will have a different header and identification name for our time code.

At the interface between the MPEG and ATM systems, there will have to be the capability to look up the incoming header name, recognize it as time code, look up the appropriate identification for time code in the outgoing system, and then structure a new header to wrap around our time code. At the interface between the ATM system and the third system, there will have to be a similar capability with a completely different set of look-up tables. And all of these look-up tables will have to be updated whenever there is some new data type to be carried, since we said we wanted our systems to be extensible but each assigns its own identification to the same data in its particular header scheme. This is all rather chaotic and quite burdensome.

The solution to this dilemma is a universal header system with a universally applied content identification. This would allow the same data to be identified in the same way in all transport systems without regard to the particulars of their individual transport header schemes. It requires that there be a single, worldwide mechanism for assigning content identifiers. It also requires that all standardization bodies around the world adopt the same technique for all of their packetized systems.

This is a big undertaking. In fact, a SMPTE Working Group is trying to develop just such a system. As you might imagine, given the implications this will have for systems of all kinds for decades to come, the discussion has been rather contentious. A consensus is within reach, however, on the kernel of such a system, and it is hoped that agreement can be reached before the window of opportunity closes and chaos reigns. It should be noted that all interested parties are welcome to participate in these discussions.

RASTER-BASED SIGNALS

Finally, what might we do to transport raster-based (non-compressed) signals in the future when television facilities are replete with high speed, packet switched distribution systems carrying myriad compressed signals of all types? It will still be necessary to transport and store non-compressed signals for production and post production purposes before they are ready for compression. But it may be that, because of economies of scale, packetized distribution systems will be significantly less expensive than the dedicated distribution schemes of today. Certainly it would be possible to maintain a separate system for raster-based images, but this could be inefficient and costly.

Why not, then, handle the raster-based images using packetization? In fact, they are already structured in a way to suggest this. Each line of the active picture area could serve as the payload of an elementary stream type of packet. Perhaps the luminance and color difference signals should be packaged separately. Some mechanism would be needed to identify which line of an image each packet represented, and packets would also have to be identified as members of the same field or frame in order to keep an image sequence sorted out. But all of this is relatively trivial to do. If the high speed, wide bandwidth networks are in place, it will be only a matter of time before someone builds such a system as described here.

As we said when we began this little miniseries, the impact of the change to packetized communication will be widespread and profound. It will affect the way images and sound are stored and recovered, transmitted and received, distributed and switched. It will also enable the inclusion of television material in the more general flow of data and communications that will result from the eventual convergence of the computer, communications, and entertainment industries. If we've done our jobs well during our little expedition, how that might happen should now be readily apparent.

PART 6

Transmission Planning

Most of the information that is publicly available concerning transmission has to do with terrestrial broadcasting. This results from the fact that the process to define the terrestrial broadcasting system under the umbrella of the FCC Advisory Committee has been an open one. At the same time, most of the work on transmission for other members of the electronic media distribution family is done in private, often in secret. It is often so obscured with non-disclosure agreements that, even if I know something about it (which I do, because it is where I make a large proportion of my living), I cannot talk about it, let alone write about it. So, the terrestrial broadcasting efforts have gotten all the attention thus far. This situation should change somewhat in the near future.

For now, though, we will look at what has been accomplished by and for the broadcasters. We begin in Chapters 19 and 20 with a couple of the earliest articles in the series (the second and third). While they are concerned with the FCC Advisory Committee selection process to choose between proposed systems, even still including the one (at that time) remaining analog system, they provide a good grounding in what the various transmission methodologies are all about. As you read these, keep in mind that the two real winners are the QAM and VSB systems. VSB was the choice of the Grand Alliance and thus the winner in the Advisory Committee, but QAM may be the bigger winner, having captured virtually all other applications short of satellites. (Satellites require a less complex modulation scheme that permits saturation of the transponders and hence does not have an amplitude component. QPSK or a variant is normally used.)

Chapter 21 is the first of two articles that looked into the real power requirements of digital transmission. It debunked the notion that terrestrial broadcasting could use vanishingly low power levels to achieve the same coverage as currently attained. This results from the combination of the FCC's intention to move broadcasting exclusively to UHF and the fact that linear increases in coverage distance require decibel increases in power.

Chapter 22 provides a complete treatment of the VSB technology that had just won the "bake-off" against QAM to land a place in the Grand Alliance system. It is a masterpiece of innovation in the use of analog technology to carry digital signals in an environment replete with NTSC interference. There are those, however, who hold to this day to the argument that QAM is theoretically just as good, only its implementation for the bake-off testing suffered.

Finally in this part, Chapter 23 revisits the matter of required transmitter power as part of its consideration of a proposal submitted to the FCC by a large number of broadcasters suggesting how channels should be pre-assigned to stations by pairing them with existing channels.

While this section is essentially devoted to broadcast transmission, much of what you need to know to consider or use this technology in other types of applications is also covered. It just isn't stated quite so directly.

19

Modulation and Channel Coding Methods

(Note to non-broadcast readers: Even though this chapter and the next concentrate on broadcast system proposals and transmitter issues, the technology discussed has been proposed as well for other media such as cable and wireless cable and is equally applicable to all media.)

Of all the many aspects of Advanced Television, the terrestrial broadcast transmission plant is probably the most fraught with challenges for system designers and operators. It offers the greatest number of different potential implementations, will have the greatest licensing and regulatory oversight, will likely be the greatest single cost element in implementation, and will likely have the highest profile exposure to public interest in the details of the implementation of ATV. It certainly may also have the largest impact on a television station's long-term ability to reach and serve its audience. And its construction will take place at a time when there will be little or no audience yet established for ATV.

Consequently, it is most important that those who will participate in terrestrial broadcasting of ATV begin the process of understanding and planning the required transmitter facilities as soon as possible. This will permit the maximum amount of time for completion of the required tasks while still meeting the FCC's deadlines. It will also permit the maximum number of possibilities to be explored, thereby yielding optimal solutions in terms of coverage, timeliness, and cost. To this end, this and the next installment will examine some of the factors that will enter into the

---

First published August, 1992. Number 2 in the series.

Table 19.1 – Proposed Terrestrial Broadcast Systems & Channel Coding

System/ Proponent	Channel Coding Designation	Basic Technology	Description
Narrow MUSE /NHK	Frequency Split Modulation	Pulse Amplitude Modulation (PAM)	PAM stream carrying digital representation of analog signal is split into segments above and below NTSC carrier frequency.
DigiCipher /General Instrument	32-QAM or 16-QAM	Quadrature Amplitude Modulation (QAM)	Constellation of specific points in amplitude and phase of carrier carry 5 (32-QAM) or 4 (16-QAM) bits of data.
DSC-HDTV /Zenith-ATT	4-VSB/2-VSB	Multi-Level Vestigial Sideband (VSB)	Vestigial sideband amplitude modulation carries data with either 2 bits and 4 levels or 1 bit and 2 levels. Pilot carrier aids signal recovery.
AD-HDTV /ATRC	Spectrally Shaped QAM	Quadrature Amplitude Modulation (QAM)	Two 32-QAM signals are positioned around NTSC carrier frequency, with higher power used for high priority data on lower carrier.
CC-DigiCipher /MIT-GI	32-QAM or 16-QAM	Quadrature Amplitude Modulation (QAM)	Constellation of specific points in amplitude and phase of carrier carry 5 (32-QAM) or 4 (16-QAM) bits of data.

design of the ATV transmitter plant. We will also doubtless return to this area from time-to-time as more information becomes available from proponents and the various testing labs.

CHANNEL CODING TECHNOLOGY

In this column last month (Chapter 1), we looked at an overview of the ATV system and described Channel Coding as the portion of the system that takes the compressed image from the Source Coding (compression) process "and converts it to a form that can be sent through the medium to be used. Thus the Channel Coding may very well be different for terrestrial broadcast, cable, satellite, wireless cable, or tape recording, even though they carry the same form of source coding. Included in this function is any modulation required. The output signal usually has analog characteristics that embody the digital information."

Before looking at some of the specifics of transmitters, antennas, and the like, let us take a look at the proposed systems and how their various channel coding schemes work. There are currently five ATV systems under consideration for terrestrial broadcasting. Four are digital systems, and one is analog. The five systems use four different channel coding schemes with three basic technologies. With regard to transmission performance, as much as anything it is the channel coding schemes that differentiate the systems and can lead to differences in performance in the terrestrial broadcast environment. The five systems (in the order of their testing) and their respective channel coding schemes are related in Table 19.1.

Without going into why they do what they do (which we'll save for another time), let us take a brief look at the three basic approaches. Then we can see how the systems apply them. Finally, we can look at how the various choices impact the designs that will be required for transmitter plants.

Quadrature Amplitude Modulation (QAM)

Quadrature Amplitude Modulation (QAM — pronounced "kwäm") forms the basis for three of the four digital (five overall) systems. QAM uses a constellation of points (as viewed on an X-Y display much like a vectorscope) that have specific amplitudes and phases. For terrestrial broadcast, the QAM systems proposed use either 16 points (16-QAM) or 32 points (32-QAM) as shown in Figures 19.1a and 19.1b, respectively. Since 32 is 2^5, it represents 5 bits. Since 16 is 2^4, it represents 4 bits.

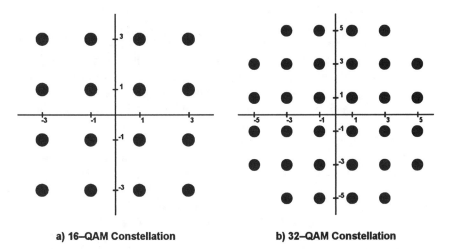

a) 16–QAM Constellation   **b) 32–QAM Constellation**

Figure 19.1 – Quadrature Amplitude Modulation (QAM) Constellations
(Shown at approximately equal power. Values indicated are
inputs to modulators to produce related constellation points.)

(256-QAM = 8 bits, 128-QAM = 7 bits, 64-QAM = 6 bits, and 8-QAM = 3 bits are also possible and may be used in non-broadcast systems.[13]) A QAM signal can be generated by suitably modulating with multi-level signals two carriers in quadrature and then adding the modulated signals together. There is inherently no carrier in a QAM signal.

As in the familiar case of NTSC color bars, the QAM signal modulation moves from one point to the next, dwells there for a while, then moves on to the next point. The time that the modulated signal dwells at a single point is called the symbol period, and the data it carries is one symbol. In NTSC-color-bar terms, we could think of this as the bar width representing the symbol period and the color representing a symbol. The fact that there is a fixed symbol period for a given system implies that there is also a fixed symbol frequency that, as it turns out, determines the data rate of that system. The data rate is the symbol frequency multiplied by the number of bits per symbol (5 for 32-QAM, 4 for 16-QAM).

Multi-Level Vestigial Sideband

The Multi-Level Vestigial Sideband (VSB) method uses amplitude modulation of a carrier by data that has been converted to either 4 amplitude levels or 2 amplitude levels. The 4 levels represent 2^2, or 2 bits. The 2 levels represent 2^1, or 1 bit. Thus this method yields 2 bits per symbol for 4 levels and 1 bit per symbol for 2 levels.[14] The 2-level signal can be thought of as similar to the digits of closed captioning signals, while the 4-level signal is similar to an unmodulated stairstep signal in which the steps can come in any order. Ordinary VSB-AM would generate a carrier, but it can be suppressed or replaced with a pilot signal.

---

[13] We know now that QAM was not selected for broadcast Advanced Television, but it is planned for wide use in cable and wireless cable applications, where 64-QAM and 256-QAM are most likely to be used. A limiting case, with only one amplitude in each phase and 4 constellation points, hence carrying 2 bits per symbol, is Quadrature Phase Shift Keying (QPSK) that is used for satellite transmission, in particular Direct Broadcast Satellite (DBS).

[14] The multi-level VSB method was, of course, ultimately selected for broadcast Advanced Television use as part of the Grand Alliance system. The technique adopted is much more sophisticated than initially proposed, using a larger number of levels and providing separate broadcast and cable modes. For broadcast, 8 levels are used (8-VSB) to carry 2 bits of data per symbol; the difference between the theoretical 3 bits that can be derived from 8 levels and the actual 2 bits is used for an error correction technique called trellis coding that effectively extends the reception threshold. For cable, 16 levels are used without trellis coding to provide twice the data rate. See Chapter 22 for details.

It might seem at first that the Multi-Level VSB approach, because it has a smaller number of bits per symbol, would have a lower data rate than the QAM systems. This does not turn out to be the case, however, because a fixed channel bandwidth (such as 6 MHz) can accomodate a symbol rate for VSB modulation that is about double that of a QAM approach. Thus the maximum data rate of the Multi-Level VSB method is about the same as for QAM. (More about the significance of the word "maximum" shortly.)

Pulse Amplitude Modulation (PAM)

Pulse Amplitude Modulation (PAM) converts data, which might for instance be representative of amplitude levels of an analog signal, into amplitude levels that can modulate a carrier. If the data is packaged 8 bits at a time, 256 (2^8) levels are produced. Because this is fundamentally an analog modulation method, some form of signal synchronization capable of detection by analog circuitry is required.

There are other forms of modulation possible to carry the kinds of signals under consideration for Advanced Television. Examples are such forms as Quadrature Phase Shift Keying (QPSK) and Coded Orthogonal Frequency Division Multiplex (COFDM). Since they have not been proposed for terrestrial broadcasting but may be applied to other portions of ATV systems, we will examine them when we look at channel coding in more detail in another installment.

CHANNEL CODING METHODS

Now that we have identified the basic technologies, let us survey their applications by the proposed systems. Some of the systems have certain spectral features designed to avoid interference into critical points of existing NTSC signals. In the figures that accompany the descriptions to follow, the spectrum of each is shown together with the NTSC spectrum of current transmissions as a reference (in Figure 19.2e). We will look at the reasons for the spectral features when we go into the theory of the systems in greater depth in a future column. For now, we will concern ourselves with aspects of the systems that will impact transmitter plant design.

DigiCipher

QAM is applied in its "pure" form by two systems: General Instrument's DigiCipher and Channel Compatible DigiCipher, developed jointly by General Instrument and MIT, as the Advanced Television Alliance (ATVA). The precise characteristics of the two systems are slightly different, but their fundamental operation is the same. The carrier frequency is in the center of the 6 MHz television channel. The symbol rate is approximately 5 Msymbols/second (MS/s), and the total data rates are about 25 Mbits/second (Mb/s) for 32-QAM and 20 Mb/s for 16-QAM. Selection of 32-QAM or 16-QAM is left to the station, depending upon its particular circumstances. 32-QAM yields higher picture performance, but 16-QAM has a lower reception threshhold and hence better coverage for a given transmitter power level. 32-QAM is considered the "normal" mode of operation.

Both 32-QAM and 16-QAM have flat spectra across the 6 MHz channel. The symbol rate has been selected so that, when combined with suitable baseband pulse shaping and IF bandpass filters, the maximum data rate is achieved in the channel with moderate guard bands at the channel edges. (See Figure 19.2a.) As mentioned previously, there is no carrier generated by the QAM modulation.

AD-HDTV

A variation of the QAM system is the Spectrally Shaped QAM (SS-QAM) system proposed by the Advanced Television Research Consortium (ATRC) for the Advanced Digital HDTV (AD-HDTV) system. It uses two 32-QAM signals in the 6 MHz channel, dividing the data between them. (See Figure 19.2b.) This dual-channel approach allows the creation of some unique differences relative to the other systems.

Taking advantage of the rejection region of the Nyquist filter in NTSC receivers tuned to co-channel NTSC stations, the lower QAM channel can be transmitted at higher power. This combined with its lower symbol rate makes it a more reliable channel, able to be received at lower signal levels. Consequently, data most needed for the creation of a picture, the so-called High Priority data, are sent on the lower channel. This yields a form of graceful degradation as the signal level falls and the upper (Standard Priority) channel suffers from the "cliff effect," the sudden failure to recover the signal that is typical of digital transmission systems using large amounts of error correction. Under such conditions, a lower quality but viewable picture, together with its audio, remains.

DSC-HDTV

The Digital Spectrum Compatible HDTV (DSC-HDTV) system developed by Zenith Electronics Corporation and AT&T takes a different approach to achieving graceful degradation. As mentioned previously, the 4-VSB/2-VSB system provides for either 4-level or 2-level modulation of the carrier. In fact, both are used in a bi-rate scheme.

The 2-level signal is more robust and can be received at lower signal levels than the 4-level signal, but the 2-level signal can only carry half the data of the 4-level signal. Either 2-level or 4-level data is used in transmission, with the ratio of 2-level and 4-level data adjusted dynamically, depending upon picture content. The 2-level data carries the most important picture information plus audio, with the 4-level signal carrying the remaining data that cannot be fit within just the 2-level data to make a high quality picture.

Because the proportion of 2-level and 4-level data changes depending upon the picture being transmitted, the total data rate varies with the proportion of 4-VSB signal *vs.* 2-VSB signal that is sent for a particular image. Thus, as mentioned earlier, there is a "maximum" data rate for the channel — it is the rate when 4-VSB is sent. But the average data rate will never reach the maximum; it will lie somewhere between the maximum of 4-VSB and the minimum of 2-VSB, depending upon picture content.

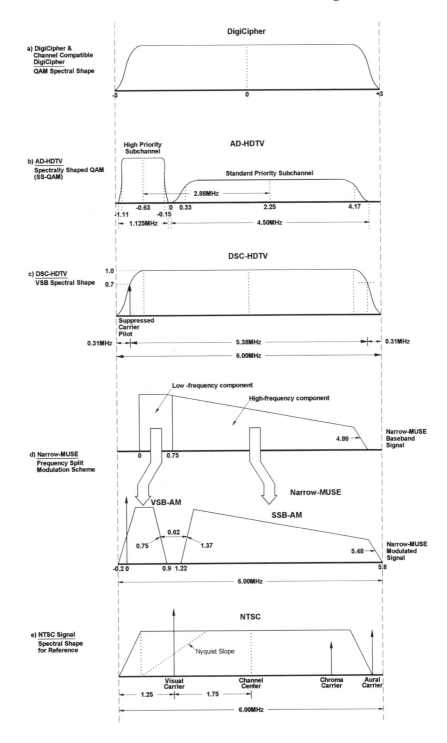

Figure 19.2 – Spectral Shapes of Proposed HDTV Modulation Systems

The DSC-HDTV system uses several other unique techniques to improve its performance in weak signal conditions or in the face of interference. One is the use of a pilot carrier to help lock the receiver demodulator, even under adverse signal and interference conditions. (See Figure 19.2c.) The pilot falls within the Nyquist filter rejection region of NTSC receivers, thereby reducing its interference. Another is the use of a time dispersion filter to spread out certain coherent pulses during transmission, thereby reducing their visibility on NTSC receivers.

Narrow-MUSE

The Narrow-MUSE system, developed by NHK, the Japan Broadcasting Corporation, splits the spectrum of the broadcast channel into two portions, above and below the NTSC picture carrier, with a gap in between them. (See Figure 19.2d.) The lower segment contains the higher energy, lower frequencies of the compressed Narrow-MUSE baseband, which is a PAM stream carrying both compressed video and data for digital audio and control signals. The lower segment is modulated using vestigial sideband amplitude modulation. The upper segment contains the lower energy, higher frequencies of the compressed baseband. It is modulated using single sideband amplitude modulation. The synchronizing signals are of the tri-level type, alternating in phase from line-to-line, and exist within the same amplitude space as the PAM modulation.

IMPLICATIONS FOR TRANSMITTER PLANTS

We will have to wait for the next installment to delve into what all this means for transmitter plants. The kinds of things we will examine include the peak-to-average ratios of the system proposals, their power requirements for coverage equivalent to NTSC, and what they mean for the sizes of transmitters, antennas, and power bills; the performance required from the various equipment elements and how that relates to current transmitter plant technology; and some of the trade-offs that might be possible in the design of a facility and their implications for coverage area. Until then, it probably would be a good idea to save this issue for reference in our discussion.

20

Practical Implications of Digital Transmission

(Note to non-broadcast readers: Even though this column concentrates on broadcast system proposals and transmitter issues, the technology discussed has been proposed as well for other media such as cable and wireless cable and is equally applicable to all media.)

Last chapter, we took a first look at some of the technology of Channel Coding – the portion of an Advanced Television system that takes the compressed image from the Source Coding (compression) process and converts it to a form that can be sent through the medium to be used – and we saw how it was proposed to be applied to terrestrial broadcasting by the several proponents of HDTV systems. But what is the practical impact of the various choices of Channel Coding approach? What will be required in transmitter facilities to handle the fundamentally digital transmissions? That will be our subject this time.

It is important to remember throughout our examination of the subject that, despite the fact that they are carrying essentially digital information, the transmission systems are all basically analog in nature. This gives them many characteristics in common. The commonality of attributes permits many of the required properties of the transmitter plant to be determined before the selection of the transmission system is made. Other characteristics are quite different between the proposed systems. We will explore the effects of these similarities and differences on the design of a transmitter plant.

First published September, 1992. Number 3 in the series.

AVOIDING NTSC INTERFERENCE

Since the stated objective of the FCC is to give each existing television station a second channel to use during the conversion period for HDTV transmission, it will be necessary to double the number of stations on the air. This will only be possible if the minimum spacing between the new HDTV stations and the old NTSC stations can be considerably less than is required between NSTC stations alone. The shorter spacing requires that the HDTV signals both cause less interference into NTSC signals and be less subject to interference from NTSC signals than would an NTSC signal in the same location with the same coverage area. Such interference avoidance is accomplished in three principal ways.

NOISE-LIKE SIGNALS

First, in the case of the four digital HDTV proposals, the transmitted signal is made to appear noise-like when viewed on an NTSC receiver. Noise is less objectionable as interference than a coherent signal would be. To appear like noise, the data must take on a random order. Merely the process of digitizing an analog video signal imparts much of the required random nature to the digital signal. In addition, the extensive processing of the compression algorithms randomizes the data further. All of this guarantees a noise-like nature for the signals so long as no coherent data is added following the compression. This is the case for the QAM signals used in the General Instrument DigiCipher, Advanced Television Alliance Channel Compatible DigiCipher (CC-DC), and Advanced Television Research Consortium Advanced Digital HDTV (AD-HDTV) systems.

The Zenith/AT&T Digital Spectrum Compatible HDTV (DSC-HDTV) system does add a few bits of coherent data following compression as a synchronization code for data recovery. It ameliorates the effect this would have in an NTSC receiver by using a time dispersion filter to spread out the synchronization code as well as all of the other data. Once it is spread out in time, the sync data overlaps some of the adjacent, random data and takes on a random character itself. A complementary time dispersion filter in the receiver restores the pulses of data to their original shapes. The time dispersion filter in the receiver has the additional benefit of making the receiver less susceptible to interference from NTSC signals by spreading out the NTSC sync pulses that might otherwise interfere with the desired digital signals.

LOW POWER TRANSMISSION

The second method for avoiding interference to NTSC transmissions is the use of lower power in the HDTV transmissions. This is possible because the reception threshold of digital transmission, with respect to both noise and interference, is considerably lower than for NTSC. It should be noted in the following discussion that the digital signals are measured on an average basis while NTSC signal power is measured at the peak of sync. We will examine the significance of this distinction later.

Table 20.1 – Power Levels of Proposed HDTV Systems (Proponent Estimates)

System	Average Power			Peak Power		
	Lo VHF	Hi VHF	UHF	Lo VHF	Hi VHF	UHF
Narrow Muse	<-12.6 dB	<-12.6 dB	<-12.6 dB	-6 dB	-6 dB	-6 dB
DigiCipher	-18 dB	-18 dB	-13 dB	-11 dB	-11 dB	-6 dB
DSC-HDTV	-15 dB	-15 dB	-12 dB	-6 dB	-6 dB	-3 dB
AD-HDTV	-12 dB	-15 dB	-11 dB	-2 dB	-5 dB	-1 dB
CC-DigiCipher	-18 dB	-18 dB	-13 dB	-11 dB	-11 dB	-6 dB

Notes: 1. All values reference NTSC power measured at peak of sync.
2. Low VHF = 100 kW (20 dBk)
3. High VHF = 316 kW (25 dBk)
4. UHF = 5,000 kW (37 dBk)

With NTSC, the noise in the displayed picture (signal-to-noise ratio – S/N) is directly dependent upon the carrier-to-noise ratio (C/N). A C/N of greater than approximately 43 dB is needed for a noise-free picture. The Grade B contour is determined by a C/N of 28.5 dB. With digital transmission, so long as the data can be accurately recovered, a noise-free picture will result.

The threshold for data recovery varies somewhat with the system but is on the order of 30 dB lower than required for a noise-free NTSC signal or about 15 dB lower than for an NTSC signal at the Grade B contour. The power estimated by the proponents to be required for their respective systems for coverage equivalent to NTSC at the Grade B contour is shown in Table 20.1. Coverage equivalent to NTSC Grade B is generally considered to be the target for the proposed HDTV systems.

Because of the lower levels required at the receiver for signal recovery, it is possible to reduce the transmitter power. Exactly how much reduction can be accomplished depends upon how much of a fade margin must be allowed for the path. It may seem a little strange to talk of broadcast signals in terms of fade margin – it is a term usually used when dealing with microwave paths. But there is a very important characteristic of digital systems that will cause us to think in terms of fade margin.

CLIFF EFFECT

Figure 20.1 shows the Bit Error Rate (BER) that results from various signal levels for one particular system when no Forward Error Correction (FEC) is applied; it also shows the Block Error Rate after error correction. You will note that without

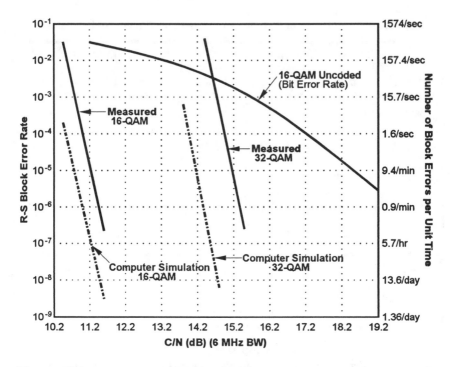

Figure 20.1 – Performance of QAM Signals With and Without Error Correction
(Reed-Solomon + Trellis coding)

error correction, the slope is gradual. But with error correction, the threshold slope
becomes extremely sharp.

The result of the error correction is that signals can be received at much lower
signal levels than would be the case without it. But when a certain point is reached,
the error correction fails, and the error rate rises very rapidly. At signal levels below
that point, the system can no longer recover the data. Video and audio disappear or
become severely distorted. The receiver likely will freeze the last image, and the
audio will mute. This is called the "cliff effect," and for good reason. A couple of the
proposed systems include multiple levels of signal robustness to provide some
amount of graceful degradation before a total loss of signal. We discussed this
approach in last chapter's installment.

How much the power of the digital HDTV systems can be reduced from NTSC
will depend on their threshold characteristics and the fade margin allowed. The fade
margin will be necessary to account for things like rain fades, airplane flutter, and
diurnal and seasonal propagation variations. The threshold levels to be used in
planning systems and the fade margins to be built into the planning factors will be
determined as part of the testing program at the Advanced Television Test Center
(ATTC) and in the deliberations of the FCC Advisory Committee on Advanced
Television Service (ACATS). A balance will be sought in this work between

avoiding abrupt interruptions in service to viewers within a station's service area and avoiding creation of additional interference to the viewers of nearby co-channel NTSC stations. The former calls for higher power while the latter calls for lower power.

The Narrow MUSE system, the only analog system among the proposals, requires essentially the same power in its sidebands as an NTSC signal to achieve the same coverage area. It achieves lower power operation by using a tri-level sync signal within the same voltage (or modulation) space as the video. This avoids a high peak power requirement for sync. Unlike the digital signals, N-MUSE average power varies with picture content, hence the less-than symbol in Table 20.1.

LEVERAGING NTSC CHARACTERISTICS

The third means for avoiding interference to NTSC stations is by taking advantage of the characteristics of the NTSC signal itself. The NTSC signal has energy concentrated at three places within its spectrum – around the video carrier, around the color subcarrier, and at the aural carrier frequencies. There is also a filter built into every NTSC receiver to lower the response to the vestigial lower sideband, thereby equalizing the frequency response across the video baseband. Three of the proposed systems make use of these attributes of NTSC to reduce their interference into it.

The drawings in the last chapter's Figures 19.2a through 19.2e show the spectra of the proposed systems and of NTSC, which is provided as a reference. The drawings are scaled to the same horizontal dimensions, so that we can easily see the relationships between the various features of the proposed new signals and the corresponding parts of the NTSC signal with which they might interfere.

AVOIDING THE CARRIERS

Both AD-HDTV and Narrow MUSE avoid placing significant energy in the regions around the NTSC visual or aural carriers. AD-HDTV does this by using two QAM signals, one below the NTSC visual carrier and one above it. (See Figure 19.2b.) N-MUSE does this by splitting its baseband spectrum and placing part in the area below the NTSC carrier and part in the area above it. (See Figure 19.2d.) Since the human eye is most sensitive to low frequencies, in both systems, the area within a few hundred kiloHertz of the NTSC visual carrier is protected.

While the DSC-HDTV system does not specifically protect part of the NTSC spectrum on transmission, it is more sophisticated in its approach to protecting the DSC-HDTV receiver from the NTSC spectrum. It uses a comb filter to reduce energy received in the regions around the NTSC visual, color, and aural carrier frequencies. It even does adaptive switching of the filter based upon the presence or absence of interference. We'll save the details of this for a time when we are considering receivers.

SKIING THE NYQUIST SLOPE

All three of the systems just mentioned take advantage of the Nyquist slope in the NTSC receiver IF filter that compensates for the presence of both sidebands in the lower video frequencies. The Nyquist slope has a falling response with falling frequency, starting at 0.75 MHz above the visual carrier and continuing to the bottom of the channel. It is down by 6 dB at the visual carrier frequency. Thus anything in the spectrum below the visual carrier will be attenuated by over 6 dB.

The AD-HDTV system places a separate QAM signal in the lower reaches of the channel. It is the one that carries the higher priority data. Because of the NTSC receiver Nyquist filter, the power in the lower QAM signal can be boosted even more than the noise-like nature of the signal by itself would allow. Thus, as is visible in Figure 19.2b, the lower QAM signal (High Priority) is 5 dB higher in level than the upper QAM signal (Standard Priority). The greater power gives the higher priority data higher reliability.

Because it is analog, the Narrow MUSE signal has its greatest energy at low video frequencies, around the carrier, with decreasing energy at higher frequencies, further from the carrier. It takes advantage of this characteristic, in combination with the NTSC receiver filter slope, by placing its carrier and the higher energy, low video frequencies in the region below the NTSC visual carrier. This reduces the impact of an interfering signal in an NTSC receiver.

The DSC-HDTV system uses a pilot carrier to help in locking its receiver under adverse channel conditions. To help reduce any potential interference into NTSC receivers, the pilot is placed near the bottom of the channel, where it will be attenuated by the NTSC Nyquist filters.

PEAK POWER *VS.* AVERAGE POWER

Noise, by definition, is not uniform. It is composed of seemingly random amounts of energy spread across the spectrum. A noise-like signal thus cannot be really constant in any of its characteristics from moment-to-moment. It will have an average value of energy, for instance, when the average is taken over a long enough period of time, but measured over a short duration, its energy will vary widely.

It is the random nature of noise or noise-like signals that first differentiates digital signals from NTSC when considering transmission and transmitters. NTSC has a constant peak signal level or power – that at the peak of sync – and an average power that varies with image content. Digital ATV signals are exactly opposite: they have a constant average value, when measured over a long enough duration, but no regular peak value. The peak amplitude can only be measured on a statistical basis.

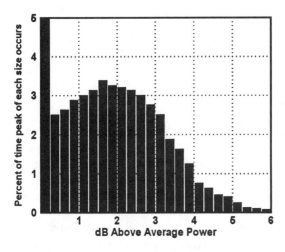

Figure 20.2 – Peak Power Distribution
for Example QAM System

Normally the peak value is expressed as the number of dB by which it exceeds the average value. The percentage of peaks that exceed a particular level above average can be measured and expressed as a statistical distribution. The effect of this is more easily seen when the distribution is charted. Figure 20.2 is an example of such a distribution for a QAM signal. The highest level of peak obtained in such measurements is used to specify the peak-to-average ratio for the system. The values for peak power (expressed relative to NTSC peak power) in Table 20.1 were derived by addition of the peak-to-average ratio to the average power (also expressed relative to NTSC peak power). The values of average power and of the peak-to-average ratio are all estimates supplied by the system proponents.

The peak-to-average ratios of the proposed systems range from 7 dB to 10 dB. The differences are caused by the characteristics of each particular system. The lowest values are found in the pure QAM systems, the highest in the AD-HDTV system, which uses two QAM signals. In AD-HDTV, if both QAM signals happen to hit peaks at the same instant, the peak powers add, resulting in a considerably higher peak than that of a single QAM channel by itself. The other systems fall in between the values for a single- and a twin- QAM channel. It is important to recognize that these peak power levels will require transmitters with peak power capabilities between 5 and 10 times their average power outputs.

The peak and average power requirements lead to specification of the characteristics of certain components in the transmission system. Peak power, for instance, determines the capacity required in power amplifiers and the voltage breakdown capability of RF hardware. Average power, on the other hand, determines the size of the power supply and the power bill. Both of these specifications will be important in acquiring transmitters in the future. The amount of peak compression that can be accepted and corrected in the error corrector of the receiver also enters into the transmission system design, but that's a matter for another time.

PRACTICAL TRANSMISSION CONSIDERATIONS

So, what's required in a transmitter plant to put all of this technology to work for us? The most important answer to this question so far is that NTSC transmission technology is good enough to carry HDTV – with careful handling. A perfect transmission system is not needed, but a good, clean system is. The system will require care in design and care in maintenance.

Any distortion of the digital signals will be translated into eye closure (another concept we will have to address on another occasion). Eye closure reduces the effective C/N ratio of the signal at the receiver. Any reduction of effective C/N reduces a station's coverage area. The kinds of distortions that are of concern – across the full 6 MHz channel, it should be noted – are: amplitude ripple, phase ripple, amplitude and phase non-linearity in active components, group delay, intermodulation characteristics, and the reflection coefficient.

Reflection coefficient is a measure of the amount of signal not radiated but reflected from an antenna back to the transmitter. Reflection coefficient has not been too much of a problem in NTSC because most facilities operate with combined transmitters, and any reflections are dumped into the reject load. Besides, for NTSC, a reflection only becomes a picture distortion (by transmitting a ghost). But for digital transmissions, a reflection translates into a reduction in C/N, which means a reduction in coverage area. This concern is exacerbated by the likelihood that early HDTV transmitters will not be parallel operations, in order to reduce their capital costs.

Intermodulation distortion in NTSC transmitters is not a particular problem because the products of the three NTSC carriers (visual, aural, color) fall outside the television channel and are suppressed by filters in the system. With digital signals, however, there will be signals across the entire channel that can intermodulate – possibly even several carriers. This will lead to in-band products that can reduce the effective C/N, again impacting coverage.

ATV transmitters are likely to be considerably simpler than those for NTSC. There will be no aural transmitters, no combiners, no notch diplexers, and none of the interlocks, cabinets, power supplies, or control circuitry associated with the aural transmitter.

One addition likely for ATV transmitters is some sort of output bandpass filter to reduce out-of-band intermodulation products to a level to be specified by the FCC. These products are noise-like to adjacent channels. They will have different effects on adjacent NTSC and digital signals. The level that will be permitted for such emissions is currently unknown; the current NTSC requirement is that they be 60 dB down in the stop band.

ATV power is currently measured as true average power – as with a calorimeter. Measuring the peak power and relating it to the average power is currently somewhat more difficult. A method has been developed by the ATTC using a frequency counter and a variable attenuator. Often, the peak power is related on a statistical basis using histograms, as discussed above and exemplified by Figure 20.2. Some amount of limiting (or compression) of peaks may be possible to permit a higher average power level with a given maximum peak level. This can be done in the amplifier by going slightly into the non-linear region, and some speculate that it can be done at IF for maximum waveform fidelity.

Test equipment for digital transmission is likely to be a mixture of familiar and new devices. Power measurement is likely to use a calorimeter to calibrate a Thru-Line® wattmeter with slightly different calibration for the digital signal *vs.* an analog signal. Peak power can be figured on a statistical relationship to the average power. Hewlett-Packard is currently developing instrumentation to measure the effect of transmission system degradation on the BER and C/N of a digital signal with an appropriate modulation format (e.g. 32-QAM or 16-QAM).

Adaptive channel equalization in the receiver is used to correct for a combination of transmitter distortions, transmitting antenna imperfections, path effects, and the receiving antenna. Only so much correction is possible for a particular implementation of the channel equalizer. As the transmitter and antenna add more imperfections, less of the total budget of correction capacity remains available for correcting for the path. This results in a lower overhead to the C/N threshold, i.e. a lower fade margin. This is okay in a strong signal area but not in a weak signal area where the margin is really needed for fades.

Most transmit antennas have some impairments in their performance – things like frequency response variations, phase variations, and such – but they are usually off the main beam of the antenna. This tends to put the impaired signals closer to the tower and hence in stronger signal areas. This fortuitous circumstance puts the impairments in places where the receiver channel equalizer can better deal with them because of the stronger signals. In weaker signal areas, the receiver is in the main lobe, where the transmitting antenna has fewer impairments and the equalizer capability can be better applied to fixing the channel. It should be noted that these comments apply to omnidirectional antennas; directional antennas may exhibit different characteristics. All effects must now be examined across the full 6 MHz rather than the critical 4.2 MHz previously required for NTSC.

Well, we've travelled all the way from theory to practice this time. Next time, we'll begin to scratch the surface of that widely used term "compression." [Now appearing as Chapter 8. If you want to follow the series in the order in which it was written, turn there. Otherwise, we'll continue with consideration of transmission by looking in the next chapter at the mantra of digital transmission – low power operation.]

21

The True Meaning of "Low Power"

Something significant happened on the way to the Special Panel meeting of the FCC Advisory Committee. In the crescendo of activity to pull together all of the information that would be needed by the Special Panel, the results of the testing at the Advanced Television Test Center (ATTC) and the results of the studies of spectrum utilization and station accommodation were released. What comes out of this release of information is a much better understanding of the nature of the digital transmissions on which future broadcasting is intended to be based. This chapter, we will examine some of the new information and look at some of its implications for digital broadcast transmission systems. While this discussion will center on broadcasting, many of the concepts will have application as well in other areas, such as cable and wireless cable.

Data was provided regarding the ATTC tests of each of the four proposed systems remaining under consideration. This data makes comparisons among the systems possible. More importantly, it begins to indicate what facilities individual television stations might require in order to achieve certain levels of coverage. This is especially instructive when viewed against some of the goals established for the allotment of channels and the criteria used to evaluate the performance of the systems. While we will look at the proposed systems in a comparative way later in this column, our principal objective will be to build the foundation for beginning to think about specific designs for particular installations. This will, in turn, allow those interested to begin to estimate costs for those facilities.

First published April, 1993. Number 10 in the series.

C/N AND INTERFERENCE LIMITATIONS

Many of the characteristics of a digital transmission system derive from the carrier-to-noise (C/N) ratio required at the receiver input in order to recover the signal after transmission. The threshold C/N ratios of the proposed systems were among the data released from the ATTC tests. Based on this, it is now possible to determine the power and antenna height required to achieve a given coverage distance or, conversely, the coverage distance that will result from a given power at a given antenna height with any of the proposed systems.

There are other factors besides C/N that control the coverage area any particular station can achieve. These are generally various forms of interference that limit the coverage in the direction of other stations on the same or adjacent channels. Such interference can be either interference caused to the nearby stations or interference received from the nearby stations. The interference caused to other stations will limit the power that can be transmitted toward those stations. The interference from those other stations will reduce a station's range below what it can achieve in other directions where there is no interference.

Giving each television broadcaster a second channel for digital transmission will require reduction of the spacing between stations to permit the effective doubling of the number of transmitters on the air. Because of this, some of the important system characteristics tested have to do with their resistance to interference. The systems have all been designed to minimize the interference from ATV into NTSC, from NTSC into ATV, and from ATV to ATV. This will help in permitting the required close spacing while decreasing the loss in coverage from interference.

There are two types of coverage limitations, then: interference-limited coverage and noise-limited coverage. While we tend to think about NTSC as noise-limited, for most stations, service is interference-limited in at least part of their noise-limited coverage areas (although in some cases, such as in calculating the Grade B contour, they are treated as being only noise-limited). The ATV allotment process has tried to match the interference-limited service contours. The data on the interference characteristics of the proposed ATV systems has been released, and we will examine interference-limited coverage in detail on another occasion. This chapter, however, we will concentrate on the noise-limited case. It is the factor that has the most to do with the coverage distance that a station can achieve in directions where there is no interference.

C/N AND POWER

The measured threshold carrier-to-noise ratios of the four proposed digital systems are given in Table 21.1. You will note that there are six entries in the table even though there are only four systems. This is because two systems, DigiCipher and Channel Compatible DigiCipher, both have 32-QAM and 16-QAM alternatives. The 16-QAM versions carry only 4/5 the amount of data that the 32-QAM signals can carry (4 bits per symbol *vs.* 5 bits per symbol), and their image quality was

Table 21.1 – C/N & Power Calculations for Proposed ATV Systems

Characteristics	Digi Cipher	DSC-HDTV	AD-HDTV	CC-DC	DigiCipher 16-QAM	CC-DC 16-QAM
C/N Threshold(dB)	15.95	15.97	18.14	15.38	11.30	11.50
Req. Field Str.F(50,90) (dBµ/m)	44.55	44.57	46.74	43.98	39.90	40.10
Field Str@ 0dBkF(50,90) (dBµ/m)	18.10	18.10	18.10	18.10	18.10	18.10
Req ATV ERP Avg. (dBk)	26.45	26.47	28.64	25.88	21.80	22.00
Req ATV ERP Avg. (kW)	441.6	443.6	731.1	387.3	151.4	158.5
Antenna Gain (dB)	14.0	14.0	14.0	14.0	14.0	14.0
Xmsn Line Loss (dB)	1.5	1.5	1.5	1.5	1.5	1.5
Req Xmtr Pwr Avg (kW)	24.95	25.06	41.31	21.88	8.55	8.95
Peak/Avg Pwr 99.9 % (dB)	6.0	7.6	6.7	6.2	5.7	6.3
Peak Xmtr Pwr 0.1% (kW)	99.33	144.21	193.22	91.21	31.77	38.18
Peak/Avg Pwr 99.0% (dB)	4.8	6.3	6.0	5.2	4.6	5.0
Peak Xmtr Pwr 1.0% (kW)	75.35	106.90	164.46	72.45	24.66	28.30

Common Receiver Planning Factors

Antenna Impedance (ohms)	75	Antenna Factor (dBm/dBµ)	-130.7
Thermal Noise (dBm)	-106.2	Receiver Line Loss (dB)	4
Noise Figure (dB)	10	Receiving Antenna Gain (dB)	10
Frequency (MHz)	615		

judged inadequate as tested. They are included here nevertheless to indicate the effects on the required C/N and power for alternate modulation schemes. This helps give a sense of the trade-offs that occur between power and the information that can be carried in a given bandwidth – in this case 6 MHz. It also helps show what might be achievable if image compression becomes good enough to work at the lower data rate that 16-QAM can provide.

Another thing to note about the C/N thresholds is the clustering of the values. Three systems (DigiCipher, DSC-HDTV, and CC-DigiCipher) group between 15 and 16 dB. The two 16-QAM systems are about 4 dB better than the lowest of the signals in the 15-16 dB range. And one system (AD-HDTV) is about 2¼ dB worse than the cluster at 15-16 dB. It is claimed by the ATRC, the proponents of the AD-HDTV system, that the threshold C/N of AD-HDTV can be bettered through the improvement of a technique called trellis coding. This is one of the items to be checked in a new round of testing at the ATTC, if separate testing of the systems actually comes to pass as authorized by the Advisory Committee. (It depends on whether or not the proponents are able to form a "grand alliance" from which a single system emerges and is tested in place of the four separate systems.)

From the threshold C/N, it is possible to work backwards toward the transmitter to determine what effective radiated power (ERP) is required to reach the receiver. When we talk about ERP in this case, we are talking about average signal power. Also involved in the whole situation is the matter of peak power. You will note that it appears in our table as well. We looked at the difference between peak and average power and at the peak/average power ratio in an earlier installment. (See Chapter 20.) Peak power is what will determine the size of the transmitter that will be required and the breakdown capacity of some components of the transmission system. Average power is what will determine coverage and the size of the power bill.

UHF SCENARIO

The FCC to date has indicated that ATV transmissions will take place in the same spectrum currently allocated to television broadcasting. This includes both the VHF and UHF bands. At the same time, the Commission has indicated a preference for a channel allotment plan that uses only the UHF band. (*Allocation* is the granting of spectrum for use by a particular service. *Allotment* is the provision of specific channels for use in particular cities. *Assignment* is the awarding of specific channels for use by particular stations.) The studies of allotment plans within the Advisory Committee have considered both possibilities: VHF/UHF and UHF-only. Since it is the worst case scenario and also the FCC's preference, we will use the UHF-only model in our examination.

In order to figure out the power that must be transmitted to reach a receiver, we have to know the characteristics of the channel and of the receiving system that will be used in the calculations. These values are called "receiver planning factors" and include such items as the thermal noise floor, the assumed receiver noise figure, the assumed antenna gain, the assumed antenna line loss, and the dipole factor. It is important to note that for factors that are frequency-related, such as the dipole factor, the values at the geometric mean of the band are used. Thus for UHF, 615 MHz (near the channel 38 visual carrier) is the frequency used in modelling. (Similarly, 69 MHz is used for low-band VHF, and 194 MHz is used for high-band VHF.)

When it is desired to know the true planning factors at specific frequencies, correction factors can be applied to the model. We will not go into the rigorous calculations in this article.

THE MEANING OF "LOW POWER"

Over the five years that Advanced Television has been under discussion, there have been frequent mention, many claims, and not a little puffery related to the idea of low power operation. At various times, systems have been described as needing transmitted power as low as "20 dB less than NTSC." This comparison becomes complicated by the difference between average power and peak power that we mentioned earlier. NTSC transmitted signal power is measured at the peak of sync, and the average power, which is lower than the peak, varies with the picture content (APL). ATV power is measured by its average, which remains essentially constant. ATV signals have peak levels that are quite a bit larger than the average power and occur randomly. Since the transmitter must be capable of developing the peak power, it is the value that determines transmitter size. The amount of compression or clipping affects the peak power; we will discuss it in a bit.

The comparisons to NTSC power are natural because that is what is now familiar as a reference. But what is it that we really want to know in figuring out what our transmitter facilities will be? Mostly, we will want to know what it will cost and how it will perform. Cost divides into capital cost – the cost of transmitter, antenna, transmission line, and installation – and operating cost – the cost of power and maintenance. Performance is largely a matter of coverage and reliability.

The ATTC results show that the ATV average power will indeed be lower than NTSC peak power for the same coverage area. In fact, the peak power of an ATV transmitter is also less than the NTSC peak power for the same coverage area. In order to help understand this, certain models have been used throughout the Advisory Committee studies, and we will use them here as a point of departure.

REAL POWER REQUIREMENTS

The model used for investigation of UHF transmission facilities is an equivalent to a nearly maximum UHF NTSC facility. Thus the station is assumed to use an antenna height of 1200 feet. It is further assumed to have a noise-limited coverage radius the same as the Grade B contour of an NTSC station transmitting 5 MW from that height – 89.7 km or 55.7 miles. Using these values and the C/N ratio, it becomes possible to work backward to the ERP required and from that to the size transmitter needed. Some of the important numbers that come out of these calculations are given in Table 21.1. Note the use of a couple values shown as F(50,90) plus a relationship. These values are based on signals being received at the stated signal strengths or higher, at 50 per cent of locations at the specified distance, 90 per cent of the time. For NTSC, many calculations are based on F(50,50) (50 per cent of locations, 50 per cent of the time) curves, but for the Grade B contour, F(50,90) values are used through a compensating factor. The difference causes a higher signal level to be used in order to allow for extra fade margin above the F(50,50) values. This is especially necessary for ATV transmission to compensate for the "cliff effect" that has been described on prior occasions in this column.

Note that a 14 dB gain (25X power gain) antenna is assumed in these computations. It is possible to use higher gain antennas, thereby reducing the size/cost of the transmitter. But this brings with it the disadvantages of a narrow vertical beam width that is normally accompanied by relatively deep nulls under the main lobe of the pattern. These nulls move as the tower sways, and 1200 foot towers inevitably sway. This is a serious matter with digital transmission, where the variations in signal strength mean that receivers near the nulls abruptly capture and then lose the signal because of the cliff effect. Higher gain antennas, especially directional versions, also have disadvantages in their performance in areas under the main beam in terms of gain and phase variations and other distortions across the channel and around the azimuths that push receivers closer to the cliff.

As can be seen, for the parameters used in the model, the average ERP ranges from 387 kW (25.88 dBk) to 731 kW (28.64 dBk) for the four proposed digital systems. It is 151 kW (21.80 dBk) or 158 kW (22.00 dBk) for the 16-QAM systems. Dividing these numbers by the antenna gains and adding back the transmission line losses (assumed to be 1.5 dB) results in transmitter average powers ranging from 22 kW to 41 kW for the standard systems. These are the power levels the power bill will be based upon. All of these values are shown in Table 21.1.

Next, it is necessary to add the peak-to-average ratio and determine the peak power capability of the transmitter. Continuing down Table 21.1, we have the peak power requirements based on two amounts of peak amplitude compression. These are the cases wherein 0.1 per cent and 1 per cent of the peaks exceed the peak power capability of the transmitter and are compressed (corresponding to the peak-to-average ratios for which 99.9 per cent and 99 per cent, respectively, of peaks fall below the measured threshold). For the 1 per cent cases, the peak transmitter powers required are between 72 and 164 kW. It should be noted that the peaks of ATV signals are much faster than the peaks of NTSC signals and thus have less energy. Because of this, some transmitter designs with a particular peak power in NTSC may be capable of somewhat higher peak powers in ATV so long as they have adequate voltage breakdown capability.

SOME ADDITIONAL EXAMPLES

Up to this point, we have been considering a single station that matches the "standard" model and looking at it with the various proposed systems applied to it. Now, let's look at some station configurations that do not match the model and see some of the implications. This time, we will use a single system proposal and vary the station conditions. To make the analysis a little easier, we will use 16 dB as a threshold C/N value. This is close to the cluster of values for three of the systems at the low end of the scale of "regular" systems. We will also use 5.6 dB as representative of the peak-to-average power ratio for 1 per cent compression. This is

Table 21.2 – Power Requirements to Match NTSC Coverage

Characteristics	Station 1	Station 2	Station 3	Station 4	Station 5	Station 6
NTSC Channel	6	7	7	11	13	30
NTSC Antenna HAAT (feet)	1,093	1,611	3,209	1,990	1,929	2,041
NTSC Grade B Distance (miles)	63.8	59.5	76.7	74.6	73.9	64.8
Req. Field Str. F(50,90) (dBµ/m)	44.6	44.6	44.6	44.6	44.6	44.6
Field Str @0dBk F(50,90) (dBµ/m)	9.9	19.4	16.2	11.4	11.5	19.3
Req. ATV ERP Avg. (dBk)	34.7	25.2	28.4	33.2	33.1	25.3
Req. ATV ERP Avg. (kW)	2,951	331	691	2,089	2,042	339
Req. Xmtr. Pwr. Avg. (kW)	166.7	18.7	39.1	118.0	115.4	19.1
Req. ATV ERP 1% Peak (dBk)	40.3	30.8	34.0	38.8	38.7	30.9
Req. ATV ERP 1% Peak (kW)	10,715	1,202	2,512	7,586	7,413	1,230
Req. Xmtr. Pwr. 1% Peak (kW)	605.3	67.9	141.9	428.6	418.8	69.5

the average of the four "regular" systems. The results of this examination are shown in Table 21.2.

The values in Table 21.2 are those for six existing television stations using certain assumptions. One assumption is that the ATV antenna can be placed on the same tower, at the same height as the current NTSC antenna. This is probably not possible in most cases but likely represents the best that might be achieved in terms of antenna height for the ATV transmissions of those stations. Another assumption is that it is desired to match the station's NTSC coverage with the ATV transmissions. A further assumption is that the receiver planning factors are the same as given in Table 21.1. Also assumed are the same 14 dB (25X) antenna gain and 1.5 dB transmission line loss of our earlier model. Finally, it is assumed that peak compression is allowed for 1 per cent of the peaks.

After studying the table, one thing becomes very clear: Matching at UHF the coverage of a low-frequency (VHF-Low or VHF-High) NTSC station that uses a relatively low tower is going to be a very difficult and expensive proposition. This results from the propagation advantage that VHF has over UHF. Our channel 6 example is not too unusual, either. The report of the Working Party of the Advisory Committee that studied these matters predicted a couple hundred stations with average ERPs in the range from 3.2 to 6.7 MW (yes, MegaWatts) for the systems clustered in the 15-16 dB C/N range. Add to this the factor for the peak-to-average ratio, and the power becomes astronomical. (6.7 MW + 5.6 dB yields 24.3 MW.)

The upshot of all this is that many stations will not try to duplicate their NTSC coverage at the start – certainly not from a single transmitter at the current antenna height. There are other methods and approaches that can be brought into play to deal with the issue. A major improvement in coverage with lower power can be achieved with the use of a taller tower, as can be seen from studying Table 21.2. Or, a lower fade margin can be accepted. This will effectively reduce the coverage area for reliable reception but may be acceptable when traded-off against the cost at the start when audiences are small. Yet another possibility is the use of distributed transmission techniques. (These are discussed in Part 7.) The bottom line is that careful consideration must be given to the circumstances of each station, with ingenuity engaged.

22

And the Winner Is ...
VSB ... (Maybe)

After nearly seven years, starting from some 26 proposals, with a wholesale change from analog to digital technology along the way, the FCC Advisory Committee has officially narrowed the field to a single system still under consideration. This happened at a meeting of the Technical Sub-Group of the Special Panel, on February 24, [1994] when selection of the 8-VSB transmission system proposal from Zenith, recommended by the Grand Alliance, was approved. The transmission system was the last unapproved subsystem, the others (video compression, audio compression, and transport) having been approved in October, 1993.

The selection came as the result of a "bake-off" between 32-QAM and 8-VSB modulator and demodulator (modem) sets for terrestrial broadcasting at the Advanced Television Test Center (ATTC) during January and February. Also included in the ATTC testing were modems operating at 256-QAM and 16-VSB for cable applications. The results with the higher modulation densities for cable operation were a tossup. The results for terrestrial broadcasting clearly favored the VSB implementation.

The signal-to-noise threshold parameter was a virtual tie between the two schemes for terrestrial broadcasting and favored VSB by 1.7 dB for cable. The differences for terrestrial broadcasting mostly were a couple dB here and a few dB there in areas such as co-channel and adjacent channel interference, impact of multipath signals, susceptibility to phase noise, and so on. In general these tilted toward VSB; a significant exception was resistance to airplane flutter. While the

---

First published April, 1994. Number 18 in the series.

differences of individual parameters were generally small, when applied to channel allotment issues such as the ability to accommodate all stations and the amount of resulting interference, VSB became the clear winner.

The argument has been made by some in the QAM camp that the QAM system could adopt a number of the measures that ultimately gave VSB the advantage. This points out an important aspect of the whole Advisory Committee process: Its decisions are based upon demonstrations and measurements of actual hardware implementations, not theoretical possibilities. Thus, since VSB had the better implementation, it has been selected for the Grand Alliance system, and further GA development effort will be applied to it exclusively. In fact, a number of improvements were proposed at the Technical Subgroup meeting (including some drawn from the QAM implementation), and several of them will be implemented in the hardware before further testing. (When we examine the operation of the VSB system momentarily, we will base our discussion upon the scheme expected to be included in the final system.)

COFDM STILL IN THE RUNNING?

Despite the decision by the Technical Subgroup to approve the VSB approach for testing, an effort seems likely to be undertaken by a group of broadcast interests to develop a Coded Orthogonal Frequency Division Multiplexing (COFDM) implementation tailored for use in countries with 6 MHz channels (read NTSC). The members of the Broadcasters Caucus of the Advanced Television Systems Committee (ATSC) are developing a plan to design and build a COFDM modem, funding the project at a $1 million level to start. (See Part 8 for an analysis of COFDM and a look at some of the challenges that face its implemention. See, in particular, Chapter 29 for the outcome of the broadcasters' effort.)

For COFDM to be considered by the FCC, the Broadcasters Caucus project would have to complete the design and build hardware ready for testing before the Commission acts and preferably before the Advisory Committee makes a recommendation. The current Advisory Committee master calendar calls for laboratory (objective) testing of the system between mid-November, 1994, and the end of January, 1995. Following viewing (subjective) tests, analysis of results, and the preparation of reports (and not allowing for any further delays), the final results are to be submitted at the end of March, 1995. [In reality, all these dates were delayed about 6½ months.] This means that any COFDM system must be completed and ready for test in the space of less than a year from now.

Of course, a delay from the current schedule would then be required to accommodate testing of COFDM. Whether such a delay would be allowed, how cooperative the Grand Alliance would be in providing the compression hardware and building the necessary interfaces to drive the COFDM modem, and who would pay the considerable costs of a further round of testing are just the beginning of the questions that will have to be answered if COFDM is to get serious consideration by the Advisory Committee and/or the FCC. We'll have to wait and see how all this develops.

A. 8-VSB (Terrestrial) Data Segment

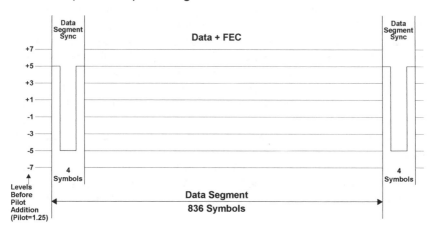

B. 16-VSB (Cable) Data Segment

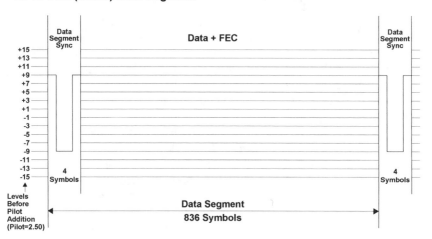

Figure 22.1 – VSB Data Segments (shown at equal power levels)

VSB OVERVIEW

The Vestigial Sideband (VSB) transmission system approved achieves the level of performance that it does through a very clever and elegant combination of techniques applied both at the transmitter and the receiver. It has two modes: one for terrestrial broadcasting and the other for cable. Fundamentally, it uses a form of pulse amplitude modulation (PAM) in which eight discrete levels are allowed for broadcast and sixteen levels are used for cable, as can be seen in Figure 22.1. This gives rise to the nomenclature 8-VSB and 16-VSB. (The levels with negative values represent a carrier phase inversion from the levels with positive values.)

During an interval called the symbol period, lasting approximately 92 nanoseconds, the PAM stream holds at a given level. This corresponds to a symbol rate of 10.762 Mega-symbols per second in the GA design. The symbol rate is chosen to have a specific relationship (684 x f_H) to the NTSC line rate in order to allow processing in the receiver that reduces the impact of NTSC co-channel interference.

Since 8 is 2^3 and 16 is 2^4, each symbol of 8-VSB represents 3 bits of data, and each symbol of 16-VSB represents 4 bits. If everything else were the same, this would mean that the 16-VSB signal would carry 4/3 times the amount of data carried by 8-VSB. We'll see a little later that everything else is not the same, and 16-VSB actually carries twice the data carried by 8-VSB. In any case, the raw data rate of 8-VSB is 32.28 Mbits/sec and of 16-VSB is 43.05 Mbits/sec. The cost for the higher data rate is a C/N threshold of approximately 27.6 dB for 16-QAM *vs.* approximately 14.8 dB for 8-VSB.

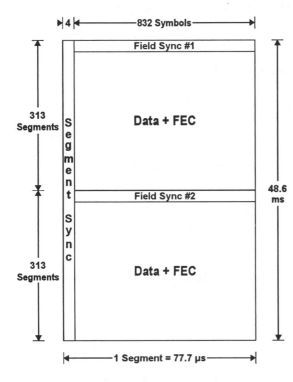

Figure 22.2 – VSB Data Frame

DATA STRUCTURE

The data stream sent to the modulator is divided into segments so that 832 symbols are included in each segment. The number of symbols is the same for 8-VSB and 16-VSB, but the data carried by the symbols is different. To this are added four symbols with a fixed pattern for synchronization purposes, as shown in both Figures 22.1 and 22.2. The segments are divided into groups of 313 to create fields, with a segment devoted to field sync at the start of each field. There are two types of fields, with the field sync signals inverted from one another to avoid creating a dc imbalance from inclusion of the fixed pattern. (Originally, the contents of the two fields were different and the sync polarity indicated which field contained a test signal. This has been removed in the latest proposals.)

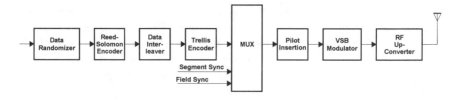

Figure 22.3 – VSB Transmitter

FORWARD ERROR CORRECTION

The value 832 for the number of symbols per segment is derived from the MPEG-2 transport stream packet length (we looked at packets and MPEG-2 transport in Chapters 16 & 17) of 188 bytes. To this is added 20 more bytes of error correcting code (ECC), using a technique called Reed-Solomon (R-S), for a total of 208 per packet. 4 times 208 is 832. The actual number of bytes and packets represented by a data segment will depend on the modulation density, but at each level there will be a fixed relationship between the packets and the data segments. In addition, as part of the processing in the 8-VSB system, another level of error correction encoding, called trellis coding, is applied. This reduces the number of payload data bits per symbol from 3 to 2 but significantly improves the robustness of the signal. After accounting for the trellis and Reed-Solomon coding overheads plus the sync signals, the net data rate of the 8-VSB system is about 19.3 Mbits/second.

As part of the process of adding the forward error correction (FEC – R-S and trellis coding), the data from the source coding process is randomized and interleaved. This has the effect of taking related data and spreading it out in time. Doing so reduces the likelihood that errors affecting a number of adjacent symbols (burst errors) will damage so much related data as to overcome the error correcting process. This helps to make the signal even more robust. The combination of error correction coding and data spreading can be seen in the transmitter block diagram of Figure 22.3. Note that the sync signals are multiplexed with the data after the data has been forward error correction encoded. This results in the sync signals being transmitted without FEC, so that they can be recovered rapidly and at very low signal levels in the receiver.

PILOT CARRIER & VSB FORMATION

Once the data structure is formed and the sync signals appended, a dc offset is added. The offset is 1.25 modulation steps for 8-VSB and 2.5 modulation steps for 16-VSB. These values result in identical amplitudes once modulation is performed, since 16-VSB modulation steps have half the amplitude of 8-VSB modulation steps. Adding the offset has the effect of slightly unbalancing the balanced mixer that provides the modulation function, thereby injecting a low level carrier signal into the output as a reference pilot.

In the implementation tested, the vestigial sideband signal was formed by generating two modulated signals in quadrature. Through complex processing of the

A. VSB

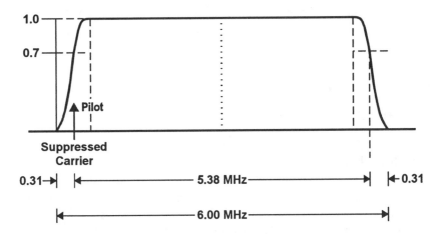

B. NTSC

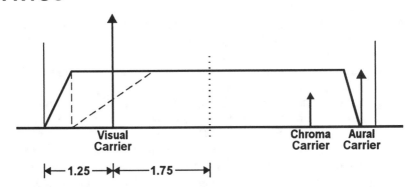

Figure 22.4 – VSB and NTSC Channel Occupancy

modulating signals, the lower sidebands were made to cancel one another when added together in the output, the upper sideband being retained. The same processing step also filtered the modulating signal to add the roll-offs required near the channel edges and to properly shape the symbols. A SAW filter then added tighter control of the energy at the channel edges and in the adjacent channels. The resulting spectrum, along with an NTSC spectrum for comparison, is shown in Figure 22.4.

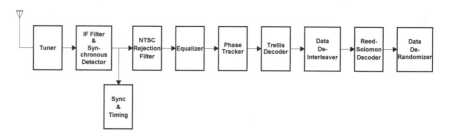

Figure 22.5 – VSB Receiver Block Diagram

RECEIVER TECHNIQUES

Just as important as formation of the signal at the transmitter is the signal recovery and complementary processing done in the receiver. In the implementation tested, shown in the simplified block diagram of Figure 22.5, a double-conversion scheme is used with the first conversion being an up-conversion to a 920 MHz first IF. The second conversion is to a center frequency of 44 MHz, the same as used in NTSC receivers. The first local oscillator (LO) is controlled by a microprocessor to set the channel. The second LO is controlled by a frequency and phase locked loop (FPLL) to fine tune the receiver and to track out any first LO phase noise. A third, fixed LO is used as the reference for synchronous detection of the received signal. Both in-phase (I) and quadrature (Q) detection of the pilot carrier are used in carrier recovery. All other processing is of the I signal only. These techniques result in the ability to recover the carrier with a 0 dB or lower signal-to-noise ratio and in the presence of significant co-channel interference.

The data segment sync signals are similarly recovered with a narrow bandwidth filter. The filter operates after A-to-D conversion by looking for a pattern match with an internal copy of the sync pattern. AGC is developed in the same circuit based upon setting the two levels of the segment sync signal to the desired digital levels in the A-D output. Because of the fixed ratio between the sync signals and the data rate, the segment sync detector also develops the 10.762 MHz reference data clock through a PLL. Once again, these circuits lead to the ability to recover the segment sync signals and to generate the data clock accurately in a 0 dB or less noise environment and where heavy interference is present.

Once data segment synchronization is accomplished and an accurate data recovery clock is generated, it is fairly simple to recover the data field synchronization. This is done by comparison of a reference copy of the field sync segment internal to the receiver with each received segment until a match is found. Since most of the field sync is sent as a two-level pseudo-random sequence, it is relatively straightforward to detect field sync with high confidence. Similar to the carrier recovery and segment sync detection, this can be done in a 0 dB noise and heavy interference environment.

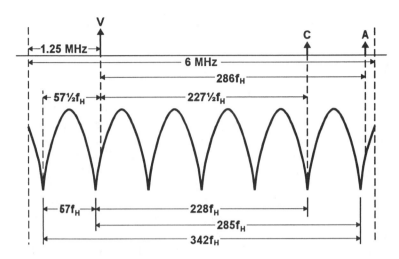

Figure 22.6 – NTSC Interference Rejection Filter

INTERFERENCE REJECTION FILTER

By virtue of the relationship between the VSB symbol rate and the NTSC line rate, it is possible to design a fairly simple comb filter in the receiver that will place nulls at the frequencies of the NTSC visual carrier, color subcarrier, and aural carrier. The filter, the response of which is shown in Figure 22.6, has seven nulls in the 6 MHz channel. The receiver compares the received signal both with and without the interference filter and chooses the path with the best results. Since the filter degrades the noise performance of the receiver by about 3 dB, it is important to use it only when it is required.

The interference rejection filter effectively creates inter-symbol interference because of the signal delay associated with it. So, a technique known as partial response detection is used to compensate for this by increasing the number of levels from 8 at the filter's input to 15 at the filter's output. The trellis decoder that comes later is designed to handle this effect.

CHANNEL EQUALIZER

A channel equalizer is required in the receiver to compensate for echoes and for tilts in the channel response. This adaptive equalizer compensates for all distortions in the path ahead of it, including transmitter and antenna distortions, path distortions, and receiver component distortions. The adaptive equalizer trains itself using the pseudo-random data sequences in the field sync intervals as a reference.

The adaptive equalizer in the implementation tested performed well on shorter, static echoes but did not do so well on longer static ghosts and airplane flutter. The next embodiment of the hardware will adopt a technique used in the equalizer for the QAM system, namely blind equalization, to improve the airplane flutter situation. The pseudo-random sequence in the field sync is also to be modified to handle longer-delayed echoes.

CARRIER OFFSET

If you look closely at Figure 22.6, you will notice that the nulls of the interference filter fall a little bit off the precise frequencies where the visual and aural carriers are located. The color subcarrier is right on the null. Since the greater amount of energy is in the visual and aural carriers, a carrier offset can be used in the ATV transmission to shift the spectrum so that the nulls fall more precisely on the carriers. With the 832 symbol segment structure, the required offset is 45.8 kHz in the direction of the upper adjacent channel.

Since the signal is attenuated significantly approaching the channel edges, and since the low level signals are shifted into an area where an NTSC receiver has substantial attenuation as a result of its Nyquist filter anyway, there are no particular upper adjacent interference problems caused by this amount of carrier/spectrum offset in the ATV transmissions.

CABLE OPERATION

For cable operation, a more benign transmission environment is assumed, and a number of the system features applied to 8-VSB for error correction and interference reduction are left out. Thus the trellis coding in the transmitter is replaced with a simple mapper to convert from serial data to multi-level symbols. In the receiver, the trellis decoder becomes a simple slicer. Similarly, the NTSC interference filter is not used in the cable receiver since strong co-channel interference should not occur on a cable system. The adaptive equalizer likely can be somewhat less capable than required for terrestrial broadcast signals, since large echoes are not likely to occur on a cable system; they can be removed at the headend. The result of these changes is a net data rate of about 38.6 Mbits/second, after accounting for the R-S coding and the synchronization signals, with a signal-to-noise threshold of about 28 dB.

A number of options are available for reception of the VSB terrestrial broadcast and cable signals. It is not difficult to imagine a single receiver able to receive either kind of signal with equal ease. Similarly, it is easy to imagine a cable receiver with a higher data rate and lower cost than that needed for terrestrial broadcast. How this will all eventually get implemented will have to play out in the marketplace. There are already a significant number of cable operators who have made a commitment to QAM systems and ordered large numbers of set top decoders. The situation is likely to get more complicated for a while before it gets better.

[The following note originally appeared with the article for June, 1994 – Chapter 16 herein. Because of its relationship to the content of this chapter, it provides better continuity located here.]

VSB REVISITED

Following publication of the article on VSB transmission in the April [1994] issue of *TV Technology*, I received a nice note from Carl Eilers of Zenith clarifying a couple of points regarding changes being made in the adaptive equalizer and its training signal, and I wanted to set the record straight. Carl points out that the upcoming change in the field sync pseudo-random sequence (from two length 255 sequences to one of length 511) is for the purpose of improving performance in general (by about 3 dB in correlation noise), not just to help with long-delayed echoes. The handling of long-delayed echoes is strictly a receiver design issue that depends on how long a filter delay is built into the adaptive equalizer. As mentioned in the article, blind equalization is being implemented to improve performance under conditions of airplane flutter.

23

The Proposed Channel Pairing Plan

On Friday, the 13th of January, [1995,] a group of 90 broadcast organizations submitted to the FCC a proposed plan for the allotment and assignment of channels in the broadcast Advanced Television Service. The submission was one of the most significant recent events in the development of digital broadcasting in the United States. This chapter we will examine the proposal and what it means for the broadcast industry and for individual television stations. (It will be left to the reader to decide whether the date itself has any significance.)

The 90 broadcast organizations included the four major commercial networks and PBS, a number of industry and trade organizations such as NAB, MSTV, and INTV, many of the major broadcast group owners including (just to pick a few) A.H. Belo, Allbritton, Bonneville, Cosmos, Cox, Gannett, Hubbard, Jefferson-Pilot, Kelly, King, Meredith, McGraw-Hill, New York Times, Outlet, Post-Newsweek, Providence Journal, Tribune, Univision, and Westinghouse, and a number of individual television stations.

With the FCC Advisory Committee scheduled to complete its work later this year, putting in place the method by which channels will be assigned to stations could quickly become a gating item in the implementation of ATV. This makes it important that decisions be reached quickly on how allotments and assignments will be made. Consequently, we can expect to see this subject get a lot of attention in the months ahead.

First published April, 1995. Number 28 in the series.

ALLOCATION, ALLOTMENT, AND ASSIGNMENT

The process of providing spectrum for the use of a television station involves three stages. The first stage is called *allocation* and is the setting aside of a portion of the spectrum for the use of a given service, in this case the Advanced Television Service. This has already been accomplished in that the FCC has indicated that the Advanced Television Service will be accommodated within the existing UHF (and, if necessary, VHF) television band(s).

The second stage, *allotment*, is the process of deciding which channels can be used in which cities. Allotment involves a tremendous amount of calculation to try out each channel in each area to see if it might be usable. The magnitude of the problem to be solved can only begin to be imagined when one thinks of the fact that the number of television transmitters must be doubled in relatively short order. The calculations use propagation prediction methods to estimate desired-to-undesired (D/U) signal ratios between existing and new, analog and digital stations. The D/U ratios obtained are then compared with the planning factors, developed in laboratory testing of the transmission system, that indicate what levels of interference between the two types of services are acceptable.

Finally, *assignment*, is the process of determining which specific channel among those allotted to a market will actually be used by each station. There are a number of possible methods for making the assignments, and these constitute some of the most significant differences between assignment proposals. The FCC suggested several possibilities in a Further Notice of Proposed Rulemaking (NPRM) a couple years ago. The 90 broadcasters have suggested a completely different approach to assignment.

FCC PLANS

The FCC has dealt with the allotment and the assignment issues on separate occasions in the ATV proceeding. In one Further Notice, the Commission published an example Table of Allotments that showed the results of running a computer program developed for the purpose. The program used more modern propagation analysis methods than are embodied in the current broadcast rules and was developed jointly with the participants in Planning Subcommittee Working Party 3 (PS/WP-3) on Spectrum Utilization and Alternatives of the FCC Advisory Committee.

The example table was intended to show how the computer model worked and to allow the industry to comment on the methods that had been used. The table was based on the FCC database frozen in its 1992 state to permit comparisons of different models without the complications of accounting for changes in the configurations of stations. The database included all then-current licensees, approved construction permits, and pending applications. The table produced was never intended as a final schedule of channels to be used in the cities listed but did serve to let many stations look at possible configurations they might be able to utilize.

In another Further Notice, the Commission offered several ideas on how it might handle the assignment process. It assumed that the allotment process would be completed first (most likely at the time of promulgation of rules establishing the ATV service) and that channel assignment would come later. The methods proposed for assignment included random pairing of channels, a lottery approach, and first-come-first-served selection by stations within a market. In the first-come-first-served method, the FCC hoped to provide a "regulatory incentive" to get broadcasters to move forward quickly in implementing ATV. (In reality, it might provide some incentive for the filing of applications but not necessarily for the early completion of implementation.)

BROADCASTER APPROACH

The 90 broadcasters have suggested that the allotment and assignment functions should be handled together and that they should be essentially completed before the service is inaugurated. In particular, they suggest that the pairing of ATV channels with NTSC channels should be done in advance of the initiation of the service and should be done by the same computer program that figures out the allotments.

The document submitted by the 90 broadcasters contends that much more efficient use of the spectrum can be achieved, along with many other benefits, through channel pairing in advance. It especially avoids the delays that would come from making assignments after the fact. It also avoids many of the legal battles likely between stations if there is an assignment process involving contention between stations for channel assignments.

The proposal assumes that ATV transmitters will be collocated with the paired NTSC transmitter and also assumes the same height for the transmitting antenna as is currently in use for NTSC. Of course, this is a practical impossibility for most stations, but it serves well for planning purposes. Provision is made in the broadcasters' plan for adjusting to the real situation later in the assignment or application processes.

Replication

The objective of the method proposed by the broadcasters is the replication of the NTSC service area with the ATV signals. This can be seen in Table 1, which is an extract of the example table of paired channels included in the document submitted to the Commission. Table 1 includes the twelve television stations currently licensed in the New York metropolitan area.

As can be seen, the example table includes the service areas, in square kilometers, calculated for the existing NTSC transmitters, using the FCC 1992 database, and the populations in those service areas. It gives a specific power level (effective radiated power – ERP) for each paired ATV transmitter and the resulting ATV service area if that power level is transmitted from the same site at the same height at which the NTSC transmitter and antenna are located. Also shown is the percentage of replication achieved by the values in the table.

While the objective is replication of service area, the actual power level that can be transmitted is limited by interference that will be created by the new ATV transmitters into both existing NTSC signals and new ATV signals. Thus, for many stations, replication less than 100 per cent is achieved, as is the case for all of the metro New York stations.

The interference received by each new ATV facility and the new interference received by each existing NTSC operation, along with the percentages of the populations affected by that interference, are given in the table. For the ATV operations, the proportion of the noise-limited service area receiving interference is shown as a percentage, with the proportion of the population in the interference region given as a percentage of the total population in the noise-limited service area. Since most NTSC signals already receive some amount of interference from other NTSC stations, the interference received by current NTSC operations from new ATV signals is given as the proportion of the service area within each station's Grade B signal area (not FCC contour – see comments below on the Propagation Model) that receives new interference. The population affected is expressed as the percentage of the total population within the Grade B signal area that is affected by that new interference.

Maximization

Beyond replication of service areas, another objective of the broadcasters' proposal is to maximize the service areas of all stations in each market, up to that of the largest current service area in the market. Thus it is anticipated that stations that currently have smaller service areas will be able to increase the sizes of their service areas, with the goal of matching the service areas of the largest stations in their markets.

To the extent possible, such maximization will be permitted from the beginning. In some proportion of cases, full maximization will not be possible initially because of interference it would cause. In most of those situations, full maximization would become possible once NTSC transmitters are turned off, if, as the FCC proposes, this eventually happens.

Public Policy Benefits

The proposal from the 90 broadcasters claims a number of public policy benefits for the approach proposed. Among these, delivering "maximum television coverage with minimum interference using spectrum to full efficiency" is foremost. This is said to be achieved through a technically driven approach that "yield[s] a table that, like a well-packed trunk, best accommodates all users in the space available."

Administrative efficiency and cost minimization are said to result from pre-assigning channels in a licensee-neutral manner that will "minimize disputes and streamline a potentially cumbersome administrative process." This comes from the pre-acceptance of the proposed process by the vast majority of stations represented among the signatories to the proposal, from telescoping the allotment and assignment functions into a single step completed before initiation of the ATV service, and from eliminating all of the procedural overhead necessary for such "administratively intensive methods" as lotteries and the first-come-first-served approach.

Enfranchisement of viewers through maintenance of existing service by broadcasters to their audiences is claimed as important. So, too, are the universal reach and local commitment of the current television service. These are all said to be best achieved through replication of existing service areas.

Finally, the proposal states that equity among television stations is promoted because "the replication/maximization method of pairing ATV channels to site specific NTSC stations has the additional virtue of reducing the coverage disparities that now exist among stations. The vast majority of stations would retain almost all of their existing coverage area[s] and many, particularly smaller NTSC stations, would receive substantially larger ATV coverage areas."

CHANNEL PAIRING PROCESS

In developing the computer model for creating a channel pairing table, three goals were sought:

"a. providing an ATV channel for each current NTSC station;

"b. providing an ATV service area that is at least comparable to the service area of the NTSC station with which it is paired and permitting stations with smaller NTSC service areas to expand their ATV service areas out to the largest service areas in the market, provided there are no adverse interference effects of doing so;

"c. minimizing the interference to existing NTSC service."

To achieve these goals, the computer model follows a sequence of steps the result of which is to provide a paired ATV channel for all of the nearly 1700 licensees, permittees, and applicants on file in 1992. The first step is to select eligible channels for each area based on the goals and interference and other criteria. Both VHF and UHF channels are included, as it was found to be impossible to achieve all the objectives using only the UHF band. (Nearly 70 VHF channels are finally included in the example table submitted with the proposal. The earlier FCC table included 17.)

Included in the list of eligible channels in each area are channels that previously were unusable because of interference they would have caused and channels that were allotted but unused. In selecting channels for an area, the model favors channels with the largest co-channel spacing to NTSC stations and other ATV allotments so as to minimize interference.

The computer model deals with the most congested markets first, progressing to less congested markets. This is a result of the fact that in the most congested markets there may be just enough eligible channels for full accommodation, while in the less congested markets there tend to be more choices of eligible channels. In New York City, for example, there are exactly twelve eligible ATV channels for the twelve existing NTSC broadcasters.

After establishing the eligible channels, the model matches each existing NTSC channel with a paired ATV channel selected from the list of eligible channels using a series of priorities. The first priority is to look for ATV channels adjacent to NTSC channels in the same market and to pair them if there will be no adjacent channel interference caused to a nearby market by the new ATV operation. This assumes that there will be exact collocation of the paired adjacent NTSC and ATV transmitters, since the resulting adjacent channel desired-to-undesired (D/U) ratio throughout the coverage area will be constant and within acceptable limits as determined by testing of the Grand Alliance transmission system.

Once adjacent channel possibilities are handled, the program tries to provide the highest replication percentage through selecting ATV channels from the list of eligibles to pair with NTSC channels. It does this by first adjusting ATV power using NTSC antenna height to match the ATV noise-limited contour with the NTSC interference-free Grade B contour of the station. It then makes a selection from the list of available channels based on the best match to the existing NTSC service area.

Because of the replication process, stations currently operating with less than maximum NTSC facilities likely will be matched with ATV facilities that produce service areas smaller than those of maximum-facility NTSC operations. This is expected to be compensated later by application of the maximization principal, allowing these stations to increase their coverage by increasing power or height or a combination, up to the largest service area in the market, so long as new interference is not caused to existing NTSC or new ATV stations.

Where NTSC stations are exactly collocated on the same tower, the program builds a pool of eligible ATV channels that provide the greatest replication for the entire group of stations. Then it matches NTSC and ATV channels in the way that maximizes replication and optimizes coverage for the group of stations.

PROPAGATION MODEL

In evaluating the coverage of existing NTSC stations and calculating the replication of new ATV facilities, the computer model adopts a number of techniques that are improvements over methods that historically have been used to predict television station coverage. The improved methods have been incorporated into the model with the participation of the FCC engineering staff and are therefore expected to be acceptable to the Commission when it finally comes time to accept an assignment methodology.

The principal improvement is the use of the Longley-Rice method of calculating propagation. The Longley-Rice technique takes into account terrain effects and other local conditions that cause both gains and losses in signal levels, and it considers predicted interference similarly affected by terrain and other conditions. Longley-Rice also takes into account the differences in propagation that occur across the various television bands, rather than considering them as homogeneous in their propagation performance. It includes these differences in calculating both coverage and interference.

The model calculates NTSC signal levels based on the directional patterns of existing NTSC stations as specified in the FCC engineering database. It also assumes that the same directional patterns will be used for ATV operations and calculates ATV service areas based on those patterns.

In addition to the improvements described, a number of other factors have been considered. These include such matters as not allotting or assigning either channel 3 or channel 4 in areas where the other is in use because of the general use of these adjacent channels by cable set top boxes and consumer VCRs, the need to provide protection to land mobile shared use of television channels, and the need to provide protection to existing Canadian and Mexican NTSC stations and to cross-border ATV stations once their locations and facilities are known.

FUTURE CHANGES

The broadcasters' proposal recognizes the possible need to make changes to the model and the likely need to make changes to the table that results from it. Changes to the model would be for the purpose of improving how the model does the channel allotment and assignment. For example, methods for determining eligible channels might be modified, or priorities might be adjusted in selecting channels for pairing. Changes of this sort would impact the entire table. The result would be the need to rerun the program for the whole country, a process that takes 17 days around the clock on high power computers. (There are approximately 67^{1700} possible solutions to the channel pairing problem.)

Changes to the table itself will certainly be needed for a number of reasons. First, there have been changes in the facilities of a number of stations during the time since the database was frozen for development purposes in 1992. A good example is included in Table 23.1. WNYE on channel 25 is shown with an elevation of 177 meters (580 feet) and at low power, which was the situation at its old location at the Brooklyn Technical High School. In reality, since just about the time the FCC database was frozen, it has been at the Empire State Building at an elevation on the order of 425 meters (almost 1400 feet) and at a power on the order of 2 megawatts. This will result in a significantly different facility to achieve replication of coverage from that shown in Table 23.1.

Table 23.1 – Extract from Preliminary Allotment/Assignment Plan Tabulated Data

Call	City, State	NTSC Chnl.	ATV Chnl.	ATV Power (kW)	HAAT (meters)	ATV				NTSC				Percent Matching
						Service Area (km²)	Populat'n (000s)	Interference % N.L. Area	Populat'n Affected %	Service Area (km²)	Populat'n (000s)	New IX % N.L. Area	Populat'n Affected %	
WNJU	Linden, NJ	47	44	300.8	460.0	15172	16235	2.7	0.9	14971	16412	1.3	0.4	98.9
WNET	Newark, NJ	13	45	206.0	500.0	25655	17802	1.2	0.3	22877	17700	0.5	0.3	99.6
WHSE	Newark, NJ	68	64	105.1	436.0	19080	16669	0.7	0.1	17529	16386	5.4	1.5	99.6
WXTV	Paterson, NJ	41	40	112.6	421.0	21222	17109	0.5	0.1	20389	17131	5.5	1.7	99.8
WWOR	Secaucus, NJ	9	18	211.6	500.0	26089	17809	1.7	0.7	22629	17332	0.0	0.0	98.8
WCBS	New York, NY	2	28	406.9	482.0	27716	17999	3.8	1.3	23543	17548	0.0	0.0	97.2
WNBC	New York, NY	4	33	309.2	515.0	28512	18167	1.0	0.4	24878	17762	0.0	0.0	98.7
WNYW	New York, NY	5	36	309.2	515.0	28674	18216	0.3	0.1	24949	17743	0.1	0.0	98.2
WABC	New York, NY	7	27	223.6	491.0	25960	17842	1.0	0.3	23333	17733	0.2	0.0	98.4
WPIX	New York, NY	11	56	200.2	506.0	25743	17796	0.9	0.3	22927	17698	0.5	0.3	98.5
WNYE	New York, NY	25	24	15.7	177.0	10069	14668	1.0	0.3	10003	14779	1.7	0.6	99.4
WNYC	New York, NY	31	53	156.4	475.0	17382	16322	1.5	0.4	17073	16586	0.7	0.2	99.4

Another type of change that will be required results from the fact that the use of current transmitter locations and antenna heights (while very useful for developing a computer model) leads in many cases to ATV transmitter facilities that will be impractical to build. This typically happens when a low-band VHF station that uses a relatively low antenna is assigned a UHF channel for ATV operations and replication of the service area is attempted.

The worst case of this situation in the preliminary table supplied with the 90 broadcasters' proposal is the current channel 6 in Columbus, OH. In the table, it is paired with channel 39 at an antenna height of 162 meters (531 feet). The ATV power shown in the table is 5.231 MW ERP. Since this is average power and ATV signals have peaks 6-7 dB above the average, peak power required in this case will be 20.9 to 26.1 MW, depending upon the peak overhead allowed. Even if a very high antenna gain of 40X is assumed (not a good idea for a number of reasons, starting with tower sway), transmitter peak power output will have to be in the range of 525-650 kW – clearly impractical.

A much better solution for this situation is the use of a taller tower. At UHF, the "height gain" that results from increasing tower heights is very significant. For instance, using the FCC F(50,90) statistics that are appropriate for digital transmission to allow adequate fade margin and the ATV field strength planning factor of 43.5 dBu, doubling the height for this station (to 324 meters – 1,063 feet) yields a 10 dB reduction in power required to reach the same noise-limited contour distance. (Your intrepid correspondent is on the road as this is written, and my laptop computer is not powerful enough to run the Longley-Rice analysis. Hence the use of the simpler FCC propagation model, which should still be very close for this purpose.)

Even if a more reasonable antenna gain of 25X is used, transmitter peak power drops to the range of 84-104 kW, with an average power of 20.9 kW (disregarding transmission line losses). This is well within the realm of practicality and shows why the table will have to be changed for specific situations.

All of which says that stations can effectively use the table in the proposal to look at their own situations and to help in forming their plans for ATV operations. At the same time, it must be recognized that the table will change and assignments will be different in many cases. It might also be that the FCC will choose not to adopt the broadcasters' approach, but some sort of similar analysis clearly will be required.

PART 7

Distributed Transmission

Distributed transmission is a technique that takes advantage of the combination of some unique properties of digital signals and the fact that virtually all digital receivers will include adaptive equalizers to process those signals. It permits a radical change in the transmission model of a single, high power transmitter with a high antenna location that has been used since the beginning of television broadcasting. Instead, it allows multiple transmitters to be used with lower power levels and from lower antenna heights in something akin to a cellular arrangement.

Distributed transmission has been something of a personal favorite concept that has been a focus of a substantial amount of work for over four years. Beginning with the original studies done for the FCC Advisory Committee and continuing with work for several clients, it is rapidly approaching the realm of practicality. It may turn out to be more applicable in areas other than broadcasting than in broadcasting itself. Nevertheless, the articles were written about broadcast applications and based upon the work done for the Advisory Committee because that is an area where I can share what I know.

It is also possible to achieve distributed transmission without adaptive equalizers by using an unorthodox modulation scheme called COFDM that has been popular in Europe. It is mentioned toward the end of the second of these two chapters and will be covered in detail in the next part of this book.

24

Distributed Transmission of Digital Signals

Throughout its 60-year history, the transmission of television programming has been in analog form, with a single transmitter operating at high power and high elevations to deliver the signals over relatively wide areas. The advent of digital television transmission brings with it the opportunity to radically alter the model, to use instead multiple, low power, low elevation transmitters distributed throughout the service area. There are a number of very significant reasons why this would be a good thing to do. There are also some very good reasons why not to do it.

This chapter and next, we will look at the operational and technical opportunities and challenges posed by **distributed transmission** and by a modulation method (COFDM) claimed to be particularly useful for such operations. Distributed transmission has potential applications in broadcast television, digital audio broadcasting, wireless cable, 28 GHz, and similar "broadcast" systems. It could also have implications for cable operations.

DISTRIBUTED TRANSMISSION DEFINED

In many ways, distributed transmission can be likened to a cellular telephone system. The service area is divided into a number of cells, each with its own transmitter. The transmitter powers can be much lower than that of a single, central transmitter. Lower tower heights can also be used than when covering a large area from a single site.

First published November, 1993. Number 14 in the series.

209

The major difference between cellular telephone systems and distributed transmission is that cellular phone systems divide the spectrum into three channel groups with individual cells using only one of the groups. Cells are assigned channel groups in a pattern that assures that no adjacent cells share the same channel group. In television or radio broadcast transmission, there likely will be no additional channels available to establish such an alternating assignment pattern. Instead, there will have to be a **single frequency network** (SFN), with all transmitters on the same frequency. It is to emphasize this difference that the term "distributed transmission" is used here instead of "cellular television," or indeed "cellular radio."

What makes distributed transmission possible is the fact that digital receivers, from television sets to set top boxes, will all require a means for dealing with echoes ("ghosts," in NTSC terms). Echoes in received digital signals normally cause inter-symbol interference (ISI) that can make it impossible to properly interpret whether a received symbol represents one value or another. There are two primary methods for dealing with the echoes: (1) adaptive equalizers in the receivers, or (2) a particular form of modulation called orthogonal frequency division multiplexing (OFDM). We will look at these in more detail next chapter.

Since receivers will have the ability to handle echoes and since multiple signals arriving at a receiver from adjacent cells of a distributed transmission system will have precisely the same modulation on them, such signals can be treated as a series of echoes. The receiver will then adapt to the "echo" environment, extracting the correct data from the ISI-laden received signals. It is this echo elimination that really makes the distributed transmission concept possible.

ADVANTAGES OF DISTRIBUTED TRANSMISSION

The farther a signal is transmitted, the more expensive becomes any marginal increase in coverage area. Put another way, the last mile of coverage is considerably more expensive to achieve than the first mile of coverage. This effect becomes accentuated with increasing frequency. For example, with NTSC broadcast transmission, a low band VHF channel achieves Grade B coverage of 65 miles or so with 100 kW effective radiated power (ERP) at 1,000 feet, while a UHF channel takes 5 MW ERP at 1,250 feet to go a little less than 56 miles to its Grade B contour (using the FCC $F(50,50)$ curves).

To increase the coverage of the NTSC VHF station from 65 to 70 miles would take almost double the power for a 7.7 per cent increase in Grade B coverage distance; the NTSC UHF station also would need about twice as much power to achieve 60 miles coverage, a similar 7.7 per cent increase in range. Furthermore, the UHF station would require a four-fold (6 dB) increase in power (totalling about 20 MW ERP) to increase its range to 65 miles (a 16.7 per cent increase in range). Similar effects affect the coverage of other services to an even greater extent since they are even higher in frequency.

All this makes achieving wide area coverage from a single site very expensive and very difficult. Since all of the television applications we are considering will be at UHF or above (assuming the FCC carries through with its plan to put ATV broadcasting exclusively in the UHF band), obtaining the power levels required for wide area coverage can be a daunting prospect.

Suppose instead the service area were divided into a number of overlapping cells. We might use a single central transmitter of much lower power and elevation surrounded by a moderate number of even lower power, lower height transmitters. We also might use a yet larger number of yet lower power transmitters in some form of grid pattern. Either arrangement will yield benefits with respect to interference to and from adjacent service areas as well as reducing the power required to cover the intended service area.

One can think of the coverage area of a transmitter as being surrounded by an annular ring of interference. The larger the coverage area, the larger the interference zone. With a single, powerful, high transmitter, the interference zone can be quite large. When less powerful, lower height transmitters are used, the interference zones around them might have about the same proportionate sizes relative to the coverage areas of the transmitters, but the absolute distances covered by the interference will be considerably smaller. Then the service area can be extended through combining the coverage areas of the several transmitters, but the interference zone around the service area will only extend so far as the shorter interference region of the perimeter transmitters. In fact, if adjacent service areas use the same channel, the interference region between the service areas needs to be only as wide as one cell.

For non-broadcast applications that can provide two-way communications capability such as wireless cable and 28 GHz systems, the use of a distributed transmission approach can bring additional benefits. No matter what part of the spectrum is used, there is never enough spectrum to satisfy all possible uses. By co-siting multiple receiving locations with multiple transmission points, the number of simultaneous communications can be increased through frequency reuse. This increases spectrum efficiency and capacity. As a practical matter, it also helps when the paths in the two sides of a communications circuit are symmetrical.

DISADVANTAGES OF DISTRIBUTED TRANSMISSION

As with so many things, the advantages of distributed transmission are counterbalanced by a number of disadvantages. First there is the need to build and maintain all of the transmitter facilities. For broadcast facilities, when a particular model is used, the equipment cost turns out to be about the same as the cost to build a single, central, high power transmitter plant. To that must be added the cost of site acquisition, however, which can include considerable legal costs when a NIMBY ("Not In My Back Yard") challenge is mounted.

Maintenance costs have been estimated to be about the same as for a single site when a small number of transmitters is used. This is because the lower power transmitters are more likely to be solid state and to have higher reliability as a consequence of their lower power operation. Nevertheless, a mobile maintenance facility would have to be developed and operated. This includes necessary staffing. Where the outlying transmitters relay the signals from the inner transmitters (about which more next chapter), redundancy becomes a more important factor since an inner failure takes down more of the system.

One of the largest disadvantages of distributed transmission is the need to deliver the signal to each of the transmitters. Depending upon the method chosen for signal delivery to the transmitters (to be discussed later), either an on-channel repeater system must be built or a separate delivery system must be provided. On-channel repeater operation with moderately high powers is problematic at best. Regarding a separate delivery system, a study for the FCC Advisory Committee showed that, without such a system, the costs for a single transmitter and for distributed transmission were about equal. However, the added expense of separate delivery made the cost of distributed transmission unacceptable. Whether this will be true for any particular system will depend on the system's specific design and the delivery resources available to it. The devil is in the details, as they say, and we will look at some of those details next chapter.

Another disadvantage of distributed transmission occurs when all of the channels to be received are not collocated. As we will see when we discuss adaptive equalizers next chapter, if standard modulation schemes are used, a directional antenna will be required at a large proportion of locations. If not all broadcast stations in a market use distributed transmission, or if they use markedly different transmitting locations, then the directional antennas will have to be re-aimed when changing from one station to another. This takes significant amounts of time when changing channels and would tend to discourage viewers from watching channels that were not clustered together at either a common antenna farm or common distributed transmission points. Of course for delivery media other than broadcast, such as wireless cable and 28 GHz, where all the channels are provided by a single entity, this would not be a problem.

DISTRIBUTED TRANSMISSION MODELS

There are two basic models that are used to describe distributed transmission systems. They are the "large cell" model and the "small cell" model. (See Figures 24.1 and 24.2.) The large cell model assumes a central transmitter surrounded by a single ring of four to eight relatively large additional transmitters. The small cell model assumes either several rings of transmitters or some type of grid pattern, with transmitters numbering in the tens or hundreds.

The choice of which model to use depends largely on what is to be done with the system. In the case of a broadcast application, where there is no spectrum reuse advantage to be gained from having more transmitters, the large cell approach is more likely optimum. This reduces the cost of the overall system and in particular the cost of delivery to each of the transmitters if a separate delivery system is chosen. In non-broadcast applications where a return channel is to be provided, the use of the small cell approach makes sense by permitting greater spectrum capacity to be gained. In this case, it is assumed that the revenues from the additional service provided will more than pay for the extra costs involved.

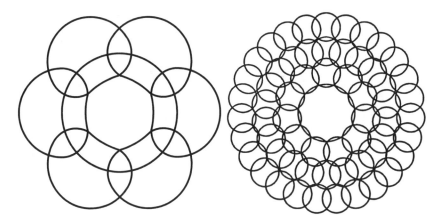

Figure 24.1 – Distributed Transmission with Large Cell Model Using Directional Antennas

Figure 24.2 – Distributed Transmission with Small Cell Model Using Omni-Directional Antennas

As has already been alluded, there are also a couple of choices regarding how to get the signals to each of the transmitters. The basic methods are through relaying from one transmitter to the next and through a separate delivery system, which can be a microwave system, for example, or fiber optics. Which of these methods to use is largely dependent on the characteristics of the adaptive equalizers to be found in receivers and whether the large cell or small cell model is used. We will look at these relationships next chapter.

Yet another characteristic to be defined is the use of omni-directional transmitting antennas or directional antennas. Once again, this choice depends on the characteristics assumed for the adaptive equalizers plus the other choices made for the model. We now have two choices in each of three categories for our system model. This gives us an eight-point matrix to examine when we go into the details in the next chapter.

As we have seen, the capabilities of the adaptive equalizer will have a tremendous impact upon what can be done in a distributed transmission system. It is thus most important to understand their relationship to system operation and the limitations they will put on system design choices. This is where we will begin in the next chapter. There we will also look at the impact of the choices of model on the operation of the system. We will look as well at some of the relative cost impacts of those choices. Finally, we will examine the possible benefits to distributed transmission operations of the COFDM modulation scheme and try to understand all the hullabaloo surrounding it.

25

The Devil's in the Details

In Chapter 24, we defined the concepts of **distributed transmission** and **single frequency networks**, and we examined some of the advantages and disadvantages of each. We also briefly discussed two models for distributed transmission systems: the "large cell" model and the "small cell" model. (See Figures 24.1 and 24.2 for drawings of theses models.) We further identified two other characteristics of distributed transmission systems, each of which offers two choices in system design: (1) the type of delivery system to the individual transmitters – direct or relayed, and (2) the type of antennas to be used – directional or omni-directional. Taken together, these three characteristics give us eight points in a three dimensional matrix for a fundamental system design. (See Figure 25.1.)

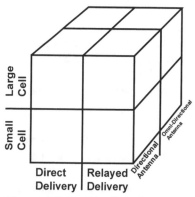

Figure 25.1 – Distributed Transmission Systems Matrix

This time we will begin by examining the technology of **adaptive equalizers** – the devices that make distributed transmission possible with standard modulation techniques. Then we will look at the interrelationships between the various possibilities in system design. Then in the next part, we will look at an alternate modulation approach (COFDM) that is claimed not to require adaptive equalizers and that may not require directional receiving antennas.

First published December, 1993. Number 15 in the series.

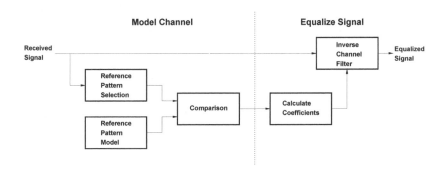

Figure 25.2 – Simplified Adaptive Equalizer

ADAPTIVE EQUALIZERS (AND GHOST CANCELLERS)

We started our discussion of distributed transmission, last chapter, with the assertion that the technique will be dependent on some method for dealing with echoes in the path from transmitter to receiver. We pointed out that this was necessary for receiving the digital signals from a single transmitter and that the signals arriving at a receiver from additional sites beyond that single transmitter would be treated as echoes. For the standard modulation methods, the device that deals with these "real" or "pseudo" echoes is the adaptive equalizer.

The adaptive equalizer treats the transmission channel as though it is some form of filter. Filters can cause variations in frequency and phase response; they can cause various time delays; and they can create signal reflections or echoes. All of these effects can result in inter-symbol interference (ISI) that makes it difficult or impossible to accurately determine what bits were sent at a particular instant.

An adaptive equalizer comprises two main parts. (See Figure 25.2.) The first of these models the channel by examining the incoming signal and comparing a received reference pattern to a copy of that reference pattern contained within the adaptive equalizer. This is similar to the Ghost Canceller Reference (GCR) recently adopted for NTSC transmission, although a digital pattern is used rather than an analog waveform. In fact, this description applies equally to the NTSC ghost cancelling process since an NTSC ghost canceller is an adaptive equalizer.

By finding the differences between the transmitted reference signal and the internal copy of the reference pattern, a description of the transmission channel in the form of a filter response can be calculated. The second part of the adaptive equalizer uses the channel model to construct a filter that is the inverse of the filter found to be in the channel. When the inverse filter is applied to the incoming signal, most or all of the effects of the channel can be removed, permitting the signal to be accurately recovered.

There are two types of filters that can be used to correct for channel distortions. They are called Finite Impulse Response (FIR) and Infinite Impulse Response (IIR)

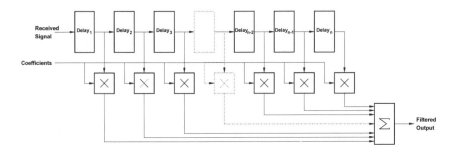

Figure 25.3 – Finite Impulse Response (FIR) Filter

filters. The FIR filter is easier to explain, but the IIR filter may be less expensive to implement and thus more likely to be used in practice. In simplified form, the FIR filter (see Figure 25.3.) uses a tapped delay line combined with a multiplier for each tap and a summing device to derive the filter characteristic needed. Control inputs to the filter are the coefficients that determine the amount of signal from each tap to be applied by the related multiplier to the summing device. Negative coefficients can be used to supply subtractive signals to the summer.

The capabilities of an adaptive equalizer are determined primarily by the length of the delay line and the number of taps/multipliers in the filter. For signals with quadrature modulation, filters are required for both the in-phase and quadrature senses of the signals. Adaptive equalizers that have been incorporated into the HDTV systems tested at the Advanced Television Test Center (ATTC) have ranged up to 26 microseconds in length (2 microseconds pre-echo and 24 microseconds post-echo) and have had as many as 280 taps. Some ghost cancellers have been built with as many as 400 taps/multipliers.

All of the delay lines, taps, multipliers, and the summing device can be implemented in either digital or analog form, but it is important to remember that the signals being processed are analog in nature while they pass through the adaptive equalizer. Any digital information is extracted later, although that digital information can be used to help control the adaptive equalizer.

Another characteristic of adaptive equalizers is that they require some difference in amplitude between the primary signal and the echoes. The difference in levels required may vary with the delay of the echoes. Thus a typical adaptive equalizer that was tested required echoes to be at least 6 dB down from the primary signal in the range from -2 to +4 microseconds and at least 12 dB down in the range from +4 to +24 microseconds.

Because the adaptive equalizer must be built into hardware at a cost that can be incorporated into consumer equipment, there must necessarily be limits on its capabilities. These limits determine such things as the maximum delay time of echoes that can be handled, the ability to handle leading echoes, the extent of frequency and phase response variations that can be handled, the difference in amplitude required between a direct signal and an echo, and the speed with which these various factors can change (as, for instance, with airplane flutter).

DISTRIBUTED TRANSMISSION SYSTEM DESIGN

When considering the design of a distributed transmission system, it helps to have a means of describing the location of a receiver relative to the nearby transmitters. This is the purpose of Figure 25.4, which describes some of the geometry of distributed transmission systems. In the drawing, three transmitters and their coverages areas are shown. The coverage areas have been set to just touch one another at a common point. Omni-directional antennas have been used in this particular case.

The signal levels at the contours represent the reception threshold plus an allowance for fade margin. This means that a receiveable signal from each transmitter most of the time travels well beyond the contour shown. It should be noted that the smaller the size of each cell, the smaller the allowance for fade margin will need to be because propagation within that cell will vary less.

Now let's look at what happens when a receiver is located at various points within the coverage areas of the three transmitters. One might think that the worst case would occur if the receiver were at point D, where equal signal levels from all three transmitters occur. This might or might not be the case, however. Three factors come into play: the timing of the transmitted signals, the adaptive equalizer characteristics, and the receiving antenna used.

If a separate delivery system is used to get the signals to the transmitters, their outputs can be synchronized with one another. In this case, the signals arriving at point D will not only be at the same level but also synchronized in time. This puts the adaptive equalizer in the part of its range where only a 6 dB difference is required between the primary signal and any echoes. Thus a directional receiving antenna would be required with the ability to differentiate between its forward lobe and any sidelobe signals by at least 6 dB. (We will call this the front-to-sidelobe-envelope ratio.) The receiving antenna can be pointed at any one of the transmitters.

If the delivery system to the transmitters is by relaying from one to the next, let's assume transmitter *a* feeds transmitters *b* and *c*. Then for a receiver located at point D and aimed at transmitter *a*, the signals from

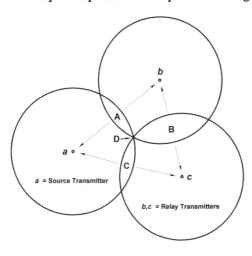

Figure 25.4 – Geometry of Distributed
Transmission Systems

transmitters b and c will arrive delayed by a bit more than the time the signal takes to propagate from a to b or c. Signal propagation is a little over 5 microseconds per mile. So, unless the cells are miniscule, the adaptive equalizer of a receiver at point D will be in the portion of its range where a 12 dB difference in levels is required between the primary signal and echoes. Thus a directional receiving antenna with a front-to-sidelobe-envelope ratio of at least 12 dB will be required.

Two things should now be apparent from these examples. First, the sizes of the cells that can be created will be totally dependent on the capabilities of the adaptive equalizers to be found in receivers. They will also be partially dependent on the receiving antennas that are available to provide the necessary discrimination between transmitters. Second and perhaps a bit more obscure, in the case of relay delivery of the signals, the use of directional transmitting antennas can help provide the signal differentiation needed. In this case, receiving antennas would always be pointed toward the nearest transmitter in the general direction of the center of the network.

If we now move the receiver away from point D, we can see what happens to the requirements for the adaptive equalizer and the receiving antenna. Different considerations apply depending upon the form of signal delivery to the transmitters and whether or not they are synchronous with one another. In the case of separate delivery and synchronous transmitter operation, moving the receiver through regions A, B, or C will keep the signals from two of the three transmitters at roughly equal levels and arrival times, but the signal from the third transmitter will begin to drop in level and increase in delay. At some point depending upon the sizes of the cells, a better receiving antenna will be required to provide the level discrimination needed by the adaptive equalizer as it begins working further out in its range.

Still in the separate delivery, synchronous transmitter timing case, as the receiver moves out of regions A, B, or C toward one of the transmitters, the antenna discrimination required will eventually drop as the signal of the nearby transmitter increases. At the same time the adaptive equalizer will begin operating further out in its range with respect to the signals from the other transmitters. In the limit, the adaptive equalizer must be able to handle signals delayed by the time related to the separation of the cell sites.

In the case of relaying the signals from one cell to the next, moving the receiver across regions A, B, or C will keep the signals from two of the three transmitters at roughly equal levels, and antenna discrimination will have to be about the same (12 dB) as at point D. Moving out of the A, B, or C region will not require more from the antenna, but it will push the adaptive equalizer further. If the receiver moves toward transmitter a, the echoes from b or c will be roughly proportional to double the time from the receiver to the particular transmitter. The antenna discrimination required will eventually drop as the signal from b or c drops and that from a increases. If the receiver moves toward b or c, then the antenna discrimination needed will drop as the receiver approaches them, but the adaptive equalizer will have to handle a <u>leading</u> echo from a that is ahead proportional to roughly double the distance to b or c. As the receiver moves beyond b or c, the adaptive equalizer will still have to process a leading echo from *a*, but it will be leading only by a time roughly proportional to the distance to *b* or *c*.

It should be quite apparent by now that all of this kind of analysis becomes very complex very quickly. There are numerous interactions between the choices made in the eight point matrix of system possibilities and the characteristics of the adaptive equalizer and the receiving antenna. It is not clear what level of adaptive equalizer capability will actually be provided in receivers or set top converters for the various forms of transmission. Notice that, in certain potential design situations, substantially more leading echo capability would be required than has hitherto been demonstrated. It seems unlikely that receiver manufacturers will burden all receivers with all the capabilities discussed in order to satisfy a few system implementations.

Distributed transmission in the form described here certainly will make it difficult for the homeowner to know what type of antenna performance is required and to make the right choices. If all of the signals to be received are not coming from a common group of sources, an antenna rotator will be needed, and analysis of each channel to be received will be required in order to make a proper equipment selection.

All of this points to the desirability of some other method if distributed transmission is to be used successfully. Ideally, such a method would use omni-directional receiving antennas or possibly even rabbit ear antennas. Ideally, too, it would permit relay delivery of the signals to transmitters. A modulation scheme which is claimed to the have the potential to do some of these things is called COFDM (Coded Orthogonal Frequency Division Multiplexing). But COFDM is so far unproven for applications such as transmission of compressed digital television signals. In the next chapter we will begin an examination of COFDM, looking into how it works and why it might be beneficial if it works as claimed.

PART 8

Coded Orthogonal Frequency Division Multiplexing

COFDM is an unorthodox, rather sophisticated modulation scheme that has been considered on-and-off for terrestrial broadcasting for over six years as of this writing. It was originally considered as the modulation approach for one of the proposed digital ATV systems but was abandoned for another method because of the time it would have taken to design and build a world class version for testing. Later, it became of personal interest because of the opportunity it offered to support distributed transmission.

Development of COFDM has been largely in Europe, with some work coming later in Japan. In Europe as this is written, it is very close to adoption by the Digital Video Broadcasting (DVB) group as the standard method for terrestrial digital transmission. A late-in-the-game effort was mounted by some North American broadcasters to gain consideration of COFDM as a replacement for the Grand Alliance VSB scheme, but it was too little, too late to succeed.

The first two chapters in this part (Chapters 26 and 27) provide technical descriptions of the workings of COFDM. Chapter 28 lays out the information obtained during a visit of technical experts to the labs where it is being developed, along with the work yet to be done (at that time) to make it practical. Finally, Chapter 29 looks at the last minute effort to substitute COFDM, the Advisory Committee's consideration of that effort, and the motivations of some of the parties on both sides of the COFDM exercise.

26

COFDM —
Dark Horse Candidate
for Digital Transmission

In our last two episodes (Chapters 24 &25), we have discussed the advantages, disadvantages, and complexities of the **distributed transmission** concept. We have seen that it has much to offer in terms of interference reduction, lower power operation, spectrum efficiency, and the like. It also has some serious disadvantages, both for the broadcaster and the consumer, when applied to "standard" modulation schemes such as QAM and VSB. It requires increased capabilities in adaptive equalizers and the use of higher performing directional antennas for reception, and it probably would work best in most cases with the use of a separate delivery system to get signals to each of the transmitters.

Suppose, instead, there were a transmission method that required neither directional antennas nor adaptive equalizers to overcome the effects of multipath propagation, that worked reliably with signals relayed from transmitter to transmitter, and that had all the advantages normally associated with distributed transmission. Suppose, in addition, that in a multi-transmitter environment the scheme allowed reception at lower signal level thresholds, that it allowed use of omni-directional (read "rabbit ear") receiving antennas, and that it cost no more in the transmission or receiving equipment than conventional methods. Such are some of the claims being made by proponents of Coded Orthogonal Frequency Division Multiplexing (COFDM), the subject

First published February, 1994. Number 16 in the series.

of our investigations this time and next, as we begin a two-part series on COFDM.

FIRST, A LITTLE HISTORY...

OFDM, the progenitor of COFDM has been around for quite a while. (We'll see what the difference is between the two in a little bit.) This modern version was first described in 1966 by the Bell Laboratories researcher who patented it, but an earlier scheme that was predictive of the concept goes all the way back to the days of the radiotelegraph. In more recent times, work has concentrated on developing algorithms that will permit high performance with implementation in relatively inexpensive integrated circuitry.

COFDM was first applied to broadcast applications in Europe as part of the work on Digital Audio Broadcasting (DAB). It was intended to provide a method for receiving digital terrestrial radio broadcasts in a moving car with only the typical vehicular monopole antenna. This meant, of course, that use of an adaptive equalizer was to be avoided, if at all possible, because of the time it takes for such a device to converge on the proper filter characteristics to correct for vagaries of the channel. By the time a calculation had been made for a given spot, the car would already have moved to a new location with a completely different set of propagation characteristics to be compensated. This was not really workable.

Another problem to be addressed by the transmission method for DAB was the wide variations in signal level that occur in a moving automobile. With the current AM and FM radio transmission systems, if signal levels drop, the signal becomes progressively noisier; with digital transmission, the signal is recovered perfectly until the threshold is reached, at which point reception abruptly stops (the "cliff effect" we have discussed several times previously in this column). Since it would be necessary to have repeater transmitters at various places in the coverage area, as, for instance, in tunnels, it was also important that simultaneous reception of several transmitters not cause destructive interference. All of this led to development of many of the techniques we shall shortly explore.

In the early days of development of the digital approaches to HDTV, consideration was given to use of COFDM by one of the system proponents. In 1990, a comparison was conducted between COFDM and a form of QAM to determine which to use in a system to be submitted to the FCC Advisory Committee for testing. A decision was reached that COFDM would take too long to develop to meet the FCC's tight timetable relative to the then unproven benefits it might offer, so the QAM method was chosen.

But COFDM keeps coming back. In Europe, it is being developed as part of three different systems for Digital Terrestrial Television Broadcasting (DTTB), which allow for both HDTV and multi-program, standard television transmission. In North America, a great deal of work has been done in Canada to define the capabilities and devise the technology. In addition, NHK (Japan

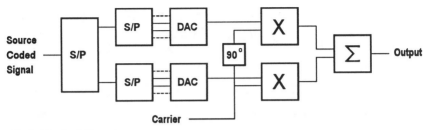

S/P = Serial-to-Parallel Converter
DAC = Digital-to-Analog Converter
X = Multiplier

Figure 26.1 – Simplified QAM Modulator

Broadcasting Corporation) has also recently acknowledged that it is building a COFDM modem.

Several submissions to the FCC on the subject led the Commission to issue instructions to the Advisory Committee in October, 1992, to study the matter and report back on the results. The Commission deferred consideration of the use of multiple low power transmitters with the then existing proposed systems until hearing back from the Advisory Committee.

Very recently (December, 1993), an Experts Group from the Advisory Committee visited the laboratories in Europe where the work is being done. Its report is expected to be completed contemporaneously with the appearance of this article. Its findings are likely to be that COFDM should continue to be studied but that the testing of the Grand Alliance system should not be held up waiting for COFDM developments. We'll discuss the findings of the Experts Group in the next chapter.

CLASSIC DIGITAL MODULATION TECHNIQUES

Consider a standard Quadrature Amplitude Modulation (QAM) signal of the sort we have discussed several times before in this column. A simplified conceptual block diagram of a QAM modulator is shown in Figure 26.1. A serial data stream carrying the source-coded signal is fed into a serial-to-parallel (S/P) converter that divides the data stream into two outputs. The two outputs feed a pair of S/P converters that, in turn, feed a pair of digital-to-analog converters (DACs). The DAC outputs modulate two carriers in quadrature that are suppressed in the modulation process (just as in NTSC subcarrier modulation).

The number of inputs to the DACs determines the modulation level of the signal. If there are two lines from the S/P converters to the DACs, the DACs put out four levels. (Since we are counting in binary, and $2^2=4$.) Four levels in each of two dimensions creates a 16 point matrix, and the result is a constellation with 16 points, or a 16-QAM signal as shown in Figure 26.2. If there were three lines to each DAC ($2^3=8$), a 64-QAM signal would result.

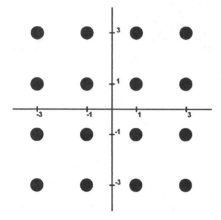

Figure 26.2 – 16-QAM Constellation

(Note that a 32-QAM signal is a special case requiring DACs with six output levels and appropriate coding to drop four points in the constellation.)

The length of time that the modulated signal spends at each point in the constellation before moving on to the next is called the symbol period. The number of symbols that can be transmitted in a given time period is primarily determined by a combination of the occupied bandwidth and filter characteristics. In a well designed system, about one symbol can be sent per second for each Hz of occupied bandwidth. In a 6 MHz channel, allowing for guard bands at channel edges, this means about 5.5 Mega-symbols per second. This results in a symbol period of approximately 182 nanoseconds.

In this discussion of symbol periods, notice that the number of points in the constellation was not involved. This allows for selection of the number of points independent of the bandwidth, with a trade-off being made against the carrier-to-noise ratio threshold. The relationship between the number of points in the constellation and the number of symbols is described in terms of bits per symbol (where a 16-point matrix represents 4 bits and a 64-point matrix carries 6 bits). When the two characteristics of the signal are combined, the result is described in terms of bits per second per Hertz of bandwidth (b/s/Hz). With 16-QAM we can achieve 4 b/s/Hz or a total bit rate of about 22 Mb/s in 5.5 MHz of occupied bandwidth. Similarly, 64-QAM yields 6 b/s/Hz or a total bit rate of about 33 Mb/s.

EFFECTS OF MULTIPATH

When a VHF, UHF, or microwave signal is propagated from a transmitter to a receiver, there frequently is more than one path that the signal follows. This leads to a primary signal and one or more echoes, which can occur either before or after the primary signal. If a point of reflection is such that it creates a secondary path for the signal that is 300 meters different (shorter or longer) from the primary path, an echo will be formed displaced one microsecond (leading or lagging) from the primary signal. Because such echo delays are long in relationship to the symbol period, they result in substantial inter-symbol interference (ISI). ISI prevents the demodulator from properly recovering the transmitted symbols.

The classic solution to ISI has been application of an equalizer that is the converse of the effective filter created by the channel propagation. Because the channel distortions cannot be predicted in advance and because they constantly change, the filter is made adaptive by sending a reference signal along with the signal payload to allow a computer to calculate the channel characteristics by comparing the reference to a model of the reference contained within the computer. The channel characterization can then be used to control a filter, updating its settings on a periodic basis. This process was described in detail in Chapter 25.

Adaptive equalizers take a finite amount of time to synchronize to the signal and extract the reference. They also require significant time between updates to sample the reference and calculate the new filter settings. These delays place limits on system performance. The time to synchronize to the reference and initially converge the filter contributes substantially to the time to accomplish a channel change. The update rate determines how well the adaptive equalizer can follow changes in the channel such as airplane flutter. While the delays involved can be expected to be reduced with improved computing capability in the receiver, they can never be totally eliminated and are likely to remain significant. As we discussed last chapter, adaptive equalizers also require substantial performance capabilities if distributed transmission is to be accommodated.

ENTER OFDM

Suppose we could find a way to reduce the symbol rate and thereby lengthen the symbol period so that most echoes represented only a small portion of the symbol period. This would confine the ISI to short intervals near the edges of the longer resulting symbol period where we could possibly find a way to ignore it. This, in turn, would allow us to recover the data without the need for adaptive equalizers and their performance limitations. This approach forms the basis of OFDM.

OFDM works, in principal, by modulating a large number of carriers (up to perhaps 1024) spaced closely together in frequency, hence the frequency division multiplexing (FDM) in the name. The symbol rate of any individual carrier is much lower than the symbol rate for a single carrier system, but the total symbol rate and hence the data rate remain essentially the same. To borrow the analogy used by Randy Hoffner in his column in *TV Technology*, it is like connecting the transmitter and receiver together using a parallel port instead of a serial port.

Conceptually, the operation of OFDM can be seen in the block diagram of Figure 26.3, an extension of Figure 26.1. The first serial-to-parallel converter is expanded to provide many outputs. Pairs of outputs connect to pairs of S/P converters and pairs of DACs. Each pair of DACs drives a pair of modulators. The carrier inputs to the modulators are derived from a frequency comb that

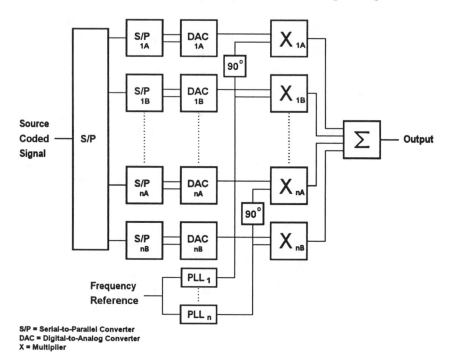

S/P = Serial-to-Parallel Converter
DAC = Digital-to-Analog Converter
X = Multiplier

Figure 26.3 – Simplified OFDM Modulator

provides equally spaced carriers throughout the channel. Once again, the modulator outputs are summed to develop the final signal.

The source coded signal arrives at the modulator at the same rate as for a single carrier approach. Because the serial data stream is divided into a large number of paths, the data within each path updates relatively slowly. The slower changes in the data result in the reduced symbol rate of each carrier. The process is reversed in the demodulator where the large number of carriers is converted back to many parallel data signals that can be combined back into the original data stream.

In practice, all the serial-to-parallel converters, digital-to-analog converters, modulators, phase shifters, and the like are not used in OFDM hardware. Instead, a mathematical process is applied to the data, causing the same effect. The mathematical process is the Discrete Fourier Transform (DFT) and is implemented using the Fast Fourier Transform (FFT) algorithm. The Fourier Transform converts between the time domain and the frequency domain and *vice versa*. This allows impulses representing the data to be converted directly into the spectrum of the output signal using the Inverse FFT (IFFT). The process at the receiver is just the opposite, using the FFT to convert the received spectrum to impulses in time.

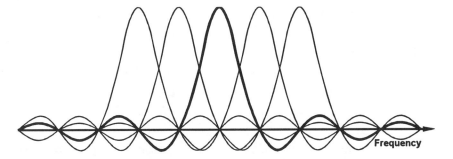

Figure 26.4 – Adjacent Orthogonal Carriers

ORTHOGONALITY

We've covered the FDM part, so where does the "O" come from? Well, when transmitting all those carriers, it is necessary to ensure that whatever data we transmit on one carrier does not affect the data recovered from a neighboring carrier. In fact, it is necessary to ensure that even the presence or absence of a neighboring carrier does not have any effect. This isolation of the carriers is called orthogonality.

Orthogonality is achieved in OFDM by setting the spacing of the carriers and the location of their sideband energy so that the energy from a given carrier reaches zero at the frequency of its neighbor on each side. Also important in maintaining orthogonality are the phase relationships between the carriers and the modulation applied to them. The amplitude and frequency characteristics of the adjacent carriers can be seen in Figure 26.4.

IGNORING ISI

Experimental observations by some of those working on OFDM systems indicate that most echoes fall within 2 microseconds of the primary signal. Thus most of the intersymbol interference will fall within a period that wide at the edges of the symbols. This can be seen in Figure 26.5. In a single-carrier system, 2 microseconds is larger than many symbol periods. In an OFDM system with, for example, 512 carriers, the symbol periods would be 93 microseconds long. 2 microseconds at the edges of such longer symbol periods represent a small portion of the energy contained within the symbol. Thus, with insertion of a 2 microsecond "guard interval" at the receiver, those 2 microseconds in which most echoes fall can be ignored without significantly reducing the reception threshold. Thus the guard interval, also shown in Figure 26.5, effectively eliminates ISI that falls within the period for which it is set, but this comes at the expense of a somewhat lower data rate through the channel and does nothing for intra-symbol interference.

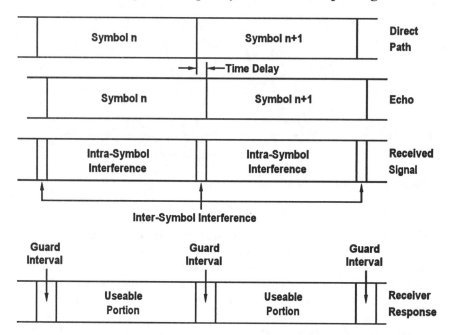

Figure 26.5 – Guard Intervals Reduce Inter-Symbol Interference

THE C WORD

Intra-symbol interference is another significant effect of echoes on recovery of data signals that is handled differently in OFDM than in single-carrier systems. In addition, the effects of co-channel signal interference must also be handled. We will go into these matters, along with some of the issues surrounding COFDM in the next chapter. Lest you be in suspense until then, the "C" in COFDM represents **coding** of the data in a manner intended to overcome both kinds of additional interference effects.

27

Coding in COFDM & Experts Group Report

In our last discussion, we began looking at the Coded Orthogonal Frequency Division Multiplex (COFDM) system that is proposed by some as the method for transmitting digital terrestrial broadcast signals. We examined the OFDM part of the system – the portion of the system that generates a large number of evenly-spaced carriers within a single channel, each carrier carrying a low data rate signal derived from the high speed signal to be transmitted. (See Chapter 26.) We also saw how "guard intervals" are used to reduce the effects of the *inter-symbol interference* that results from echoes in the transmission environment. This time we will begin with a look at how *intra-symbol interference* is handled in COFDM. Then we will see why COFDM might not be the panacea some would like to think.

It should be noted in the discussion to follow that, while the description of the system is expressed in terms of carriers being individually modulated and demodulated, in real implementations, this process is handled in digital signal processors (DSPs) that perform functions such as the Fast Fourier Transform (FFT) to process all of the carriers simultaneously and in parallel. Thus the hundreds or thousands of modulators and demodulators are reduced to a few integrated circuits.

REFLECTIONS

When the transmitted signal gets bounced around in the environment on its way to the receiver, numerous small and large echoes can be created. These would appear as ghosts in an analog transmission system such as NTSC. When they affect a

---

First published March, 1994. Number 17 in the series.

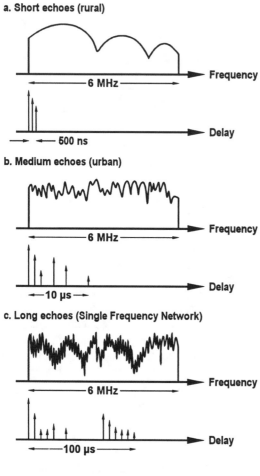

a. Short echoes (rural)

b. Medium echoes (urban)

c. Long echoes (Single Frequency Network)

Figure 27.1 – Impulse and Frequency Response
of Several
Echo Environments

digital transmission, the result is a series of notches in the frequency response of the received signal together with various phase shifts. The number and depth of the notches depend on the characteristics of the reflectors that cause the echoes. The type of terrain through which the signal propagates determines the nature of the damage done to the signal or the "channel characteristics."

Some examples of the kinds of echoes that might be found in three different environments are shown in Figure 27.1. They are described in terms of the impulse response of the channel. These include the short echoes that might be expected in a rural environment, the medium length echoes expected in an urban environment, and the longer echoes built into a distributed transmission system or "single frequency network." Also shown in Figure 27.1 are the effects of the various echoes on the channel frequency response.

Note that the few short echoes in the rural case result in a couple of widely spaced dips in the channel frequency response. The larger number of longer echoes in the urban case result in a much more ragged channel frequency response with the possibility for deeper nulls. The very long and more numerous echoes of the single frequency network can cause very deep nulls rather closely spaced in frequency. These would all be handled with an adaptive equalizer (as described in Chapter 25) when a classical digital transmission scheme such as QAM or VSB is used.

a. Constructive intra-symbol "interference" at a particular frequency

b. Destructive intra-symbol "interference" at a particular frequency

Figure 27.2 – Addition and Subtraction of In-Phase and Out-of-Phase Echoes

INTRA-SYMBOL INTERFERENCE

With the long symbol periods that come from using many parallel carriers, the echoes of the edges of transitions from one symbol to another that cause inter-symbol interference (ISI) can be significantly reduced or largely eliminated through the use of "guard intervals" as described in Chapter 26 and shown in Figure 26.5.

What remains to handle, then, is intra-symbol interference, that is interference of the echoes of a symbol with the symbol itself.

Intra-symbol interference is the summation of the direct signal with its echoes to form resultant signals that may be enhanced or destroyed by the presence of the echoes. This can be seen in Figure 27.2, where echoes that arrive at the receiver in-phase with the direct signal add to the amplitude of the sum, while echoes that arrive at the receiver out-of-phase subtract from the amplitude of the sum. Figure 27.3 should help in differentiating inter-symbol from intra-symbol interference and in seeing the potential additive effects of the summation.

a. Transmitted signal

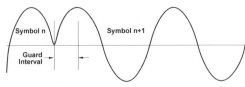

b. Received signal

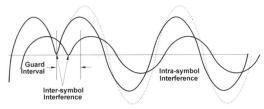

c. Channel impulse response

Figure 27.3 – Inter-Symbol and Intra-Symbol Interference

Looking again at Figure 27.1, it should be apparent that, where there are significant echoes on a path, some of the carriers essentially will be destroyed by the phase cancellation caused by the echoes at their particular frequencies. If the modulation is done in the relatively straightforward manner described in the last chapter's Figure 26.3, where each carrier carries a small, unique portion of the total data stream, the data that is carried on those carriers will be lost. This would put a heavy burden on the error correction system that processes the total data stream once it is reassembled at the receiver.

CODING

In most digital transmission systems, there are two layers of error correction, often called the "inner loop" and the "outer loop." The inner loop is designed to handle very short errors affecting only a few transmitted symbols; the outer loop is designed to handle longer errors that could damage blocks of data. The inner loop is often associated with the modulation system, such as trellis coding at the transmitter combined with soft-decision decoding at the receiver (topics for another column). The outer loop generally uses Reed-Solomon block coding and decoding.

For OFDM systems, trellis coding and soft-decision decoding of each carrier will not work because they depend on the carrier being received at a sufficient level to decode most of the data, with only occasional symbols that are misinterpreted. In OFDM systems, carriers disappear completely because of the selective filtering of the channel caused by echoes. Thus a system designed to correct for loss of symbols on an otherwise receiveable carrier is pointless.

The technique developed to correct for the loss of carriers through selective filtering or fading involves *coding* of the data before it is separated by the first parallel-to-serial converter in the modulator so that the data appears at two places in the data stream. This coding — the 'C' in COFDM — effectively makes the data appear on two (or more) carriers that are positioned in different parts of the channel but that are linked by the data they carry. This can be seen in Figure 27.4.

At the receiver, decoding can be done in such a manner that, if a particular carrier is below a

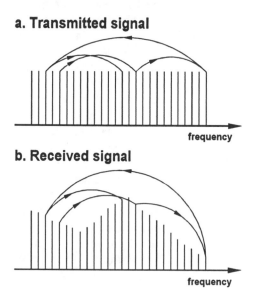

a. Transmitted signal

frequency

b. Received signal

frequency

Figure 27.4 – Coding Links Widely Separated Carriers

threshold, the data from that carrier can be ignored. The missing data is then extracted from the paired carrier that was carrying duplicate data. The system can get even more sophisticated by having the demodulator indicate to the channel decoder a measure of confidence concerning the data conveyed by each carrier, based on that carrier's level. The channel decoder can then evaluate relative confidence concerning the data from the paired carriers and make a decision about the most likely values of that data. This is essentially the same as the soft-decision decoding process in single-carrier systems.

The duplication of data on linked carriers anticipates that, if one carrier is destroyed by echoes, its linked partner will get through. It is possible, of course, for both carriers of a pair to be destroyed by echoes. Statistically, this is likely to happen infrequently if the paired carriers are randomly dispersed throughout the channel so that comb filter effects do not take out regularly spaced pairs. In order to further combat this possibility, the coding is done in such a way that the pairing is constantly changing. This further minimizes the chance that both carriers in a pair will not get through. When this does happen, however, the outer loop error correction system is expected to correct for the missing data, just as in any other dual loop error correction system.

CONSTRUCTIVE INTERFERENCE

We've looked at what happens when echoes cause destructive interference and eliminate the usefulness of some of the carriers. But what of the carriers that are increased in level by the echoes? This is called "constructive interference." To see how this works, let's consider the simple frequency comb that results from receiving two equal signals displaced in time. As shown in the example of Figure 27.5, this is essentially a direct path and a 0 dB echo with a delay of 1 microsecond.

Note that, in the frequency comb, there are frequencies where two extreme conditions appear. In one case, the signal disappears completely because the direct and echo signals are exactly out of phase and cancel one another. In the other case, however, the signals are in phase and are additive. This can result in a signal level 6 dB higher than would be obtained from a single path alone. When all the math is done, it turns out that the average signal level (power) for all the carriers across the

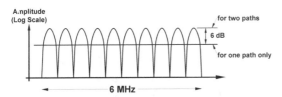

a. Impulse response

◄1 µs►

b. Frequency response

Amplitude
(Log Scale)

for two paths

6 dB

for one path only

◄――――― 6 MHz ―――――►

Figure 27.5 – Addition of Signal Power
from Echoes

channel will be 3 dB higher than if there were no echo.

This is one of the most significant potential benefits from the use of COFDM. It means that in an environment with echoes, the echoes can actually improve the reception of the signal rather than detract from it as with other modulation schemes. It also means that, where it is necessary or desirable to use multiple transmitters within a service area, the transmitters can operate cooperatively rather than interfering with one another. With classic digital modulation schemes, an adaptive equalizer is used to eliminate the effects of all but one of the received signals, and the threshold signal level required increases with increasing levels and numbers of echoes. With COFDM, in theory, just the opposite is true: the additional echoes decrease the threshold level.

When multiple transmitters are used, for instance as "gap fillers" where translators would otherwise be required, the cooperative action of the transmitters has the result that the coverage area is extended. Thus a space between two transmitters, where neither of them would deliver a signal strong enough to exceed threshold, can be filled by receiving signals from them both together, so long as the total energy received from the combination exceeds threshold.

HIGHER MODULATION LEVELS

As with so many things in nature and engineering, there is a price to be paid for the improvements that COFDM offers. That price is the fact that the coding, because it duplicates data sent through the channel, reduces the net deliverable data rate of the signal. Also included in the price is the reduction in data rate that comes from use of the guard interval that also takes time that could have otherwise carried data. To compensate for this, it is usually proposed to increase the level of the modulation from that which would have been used with a single carrier in the same circumstances. Thus if 16-QAM were used on a single carrier, 32-QAM would be required for COFDM. Similarly, 32-QAM in a single carrier implementation would lead to 64-QAM in COFDM.

Higher modulation levels (I'm tempted to use the term "modulation index" but "modulation density" is really correct) require higher received signal levels to reach threshold because of the reduced space between the points in the constellation. This has the effect of "taking back" some of the improvement that COFDM offers. The result is that the threshold level of COFDM ends up at about the same value as for single carrier systems carrying the same data rate. The real claimed improvement comes from the ability to operate well in a multipath environment or in a distributed transmission system.

CARRIER RECOVERY & LEVEL MEASUREMENT

Because COFDM will be demodulated and decoded using DSPs to handle all of the carriers at one time, it is necessary to devise a method to recover the carriers in a global manner. This must be done using hardware that costs a few dollars when systems are built by the millions so that practical application can be made to consumer uses. To date, this has only been done using techniques that require laboratory equipment that costs many orders of magnitude more than will be appropriate for consumer equipment. This is a major open issue for the potential application of COFDM.

Similarly, mechanisms must be provided to assess the level and phase of each carrier for three purposes. First, a decision must be made about whether the level of a carrier is high enough to consider the data recovered from it valid. Second, a decision must be made about the slicing levels to be used in demodulating the carrier. Another way to look at this is that some sort of automatic gain control (AGC) is required for each carrier so that different amplitudes modulated on it can be properly detected. Third, a methodology for cancelling any phase rotations of one carrier *versus* another must be provided. Techniques to perform these functions are also still in their early stages of development.

PILOT OR TRAINING SIGNALS

Among the approaches under consideration for allowing the receiver to adapt to the levels and phases of the carriers (somewhat like "adapative equalization") is inclusion in the transmission of some reference pilot carrier signals or training signals. A simple way to achieve the needed channel characterization is to insert reference carriers into the signal. Since QAM signals normally suppress their carriers, putting a small dc offset into one or both of the modulating signals will have the effect of unbalancing the modulator and transmitting some carrier. The problem with this approach is that it once again subtracts from the data space available.

Another possibility is to send some sort of training signal along with the modulation. This training signal might be of the type used for single carrier modulation or even a burst of the type used for NTSC ghost cancellation. Either of these schemes would also reduce the amount of data that could be carried by the signal. Much work remains to be done to determine the most efficient mechanism for sending to the receiver the reference it requires for channel adaptation while at the same time minimizing the data reduction required to accommodate the reference.

NO PANACEA

COFDM seems to have much to offer, but it is no panacea. A couple of examples may help to demonstrate its limitations. When multiple carriers are transmitted and each of them achieves peak power independently on a statistical basis, the total peak power can be substantially higher than that of a single carrier. The problem is very much akin to the problem of transmitting a number of single carrier signals on the same antenna (as we have discussed previously). The peak-to-average power ratio of COFDM is thus on the order of 5-6 dB higher than that for single carrier systems. This could be ameliorated somewhat through the use of carrier limiting, but at the possible expense of more difficult channel energy shaping with respect to adjacent channel interference.

Another factor is the processing time needed for channel adaptation through use of the reference pilots or training signal. The recovery of the reference and calculation of the channel characteristics will take a finite, and relatively long, time. Added to this must be the time to construct the compensating coefficients for equalization. The times involved are likely to be long enough that use with portable or mobile receivers will not be practical. This is no different from the situation with single carrier systems using adaptive equalizers, but it does mean that a significant limitation of digital systems goes unaddressed. (It should be noted that COFDM was developed for mobile applications for Digital Audio Broadcasting, and it is expected to be effective for that purpose. The difference between ATV and DAB is that the bandwidths and numbers of carriers can be much lower for DAB, and QPSK modulation can be used — at the expense of less efficient bandwidth utilization — in order to avoid the need for amplitude compensation across the channel.)

Still another consideration regarding COFDM is that with all of the various components of the signal so tightly packed together in frequency and phase, the system will be more susceptible to phase noise in any oscillators used for upconversion in transmitters and downconversion in receivers. The cost for low phase noise oscillators can be easily borne in relatively expensive transmitters, but receivers are likely to be another matter. This is especially true when one recognizes that receivers must tune over fairly wide bands and that lock-up time is important to viewer satisfaction.

VISIT TO EUROPEAN DEVELOPERS

The FCC and its Advisory Committee on Advanced Television Service (ACATS) are interested in COFDM, at least to the point of trying to determine whether it should be given serious consideration. Serious consideration would require that further delays be accepted in the process of developing and testing the Advanced Television system for the United States. To this end, in late November/early December, 1993, a Task Force on COFDM of the Transmission Expert Group of the Advisory Committee Special Panel Technical Committee (whew!) visited the laboratories in Europe where COFDM is being developed.

The Task Force's report is over 30 pages long and is summarized in the accompanying sidebar article (Chapter 28). Its bottom line is that COFDM has much to offer in improved performance in static multipath situations. This might be helpful for broadcast operations but would be of no advantage for cable. *If* that is deemed sufficiently important to delay the ACATS ATV process, then the Task Force could undertake a detailed study of COFDM and a paper design. Unfortunately, there does not currently exist any substantial body of work looking at how to implement COFDM in 6 MHz channels. (Europe uses 7 or 8 MHz channels.) Thus the Task Force would have to begin near the beginning.

With the studies accomplished, if COFDM still looked good, construction of a prototype could be sought. Estimates are that it would take 9-15 months to complete the necessary studies and build a prototype to allow optimization of the system for 6 MHz use. The prototype would then have to be tested as part of an overall ATV system. This time frame might be shortened slightly by the fact that NHK (the Japan Broadcasting Corporation) is in the process of building a COFDM system. Of course, it would be intended for 6 MHz channels. The NHK COFDM hardware was expected to be operational in September, 1994.

Recent indications are that some broadcasters are finding COFDM sufficiently attractive to proceed with an investigation as outlined by the Task Force. If this is to happen, it is likely to be decided at a meeting of the Technical Subgroup of the Special Panel on February 24 (approximately 2 weeks after this was written and 2 weeks before it was first published). We will have some comments on the outcome of that meeting in our next column.

NEW ALTERNATIVE

Fresh news from NHK is that, in addition to COFDM, its engineers are building a set of hardware for a new system called Adaptive Weighted Code Division Multiplex (AW-CDM) modulation. This system is based on the spread spectrum techniques of code division multiple access (CDMA) proposals for digital cellular telephony. Claims for this system are that it could be 2 dB better than COFDM in its reception threshold and that it might be fast enough in its channel adaptation to work for mobile applications. Watch this space for future developments.

28

Task Force Reports on COFDM

A Task Force from the Transmission Expert Group of the FCC Advisory Committee's Special Panel Technical Sub-Group visited the laboratories in Europe that are developing and advocating Coded Orthogonal Frequency Division Multiplexing (COFDM) during the week of November 28, 1993. *TV Technology* obtained an advance copy of the report and recommendation that was presented by the Task Force at a meeting of the Technical Sub-Group on February 24, 1994, after this was written.

The state of COFDM development reported by the Task Force varies from laboratory-to-laboratory in Europe. Much of the difference results from different purposes or emphases for which the COFDM systems are being developed. In each case, there is a large number of closely spaced carriers, but the modulation levels applied varied considerably. Carriers ranged from 448 (the number used for DAB work) up to 896 in current prototypes, with proposals extending up to 13,000. Some systems can operate with a small fraction of the carriers eliminated at critical points in the channel in order to reduce co-channel interference to and from nearby analog transmissions.

Modulation systems ranged from 16-QAM to 256-QAM in current prototypes, with no suggestions to exceed that number. The higher modulation levels require higher received signal-to-noise ratios in order to reach threshold – at least 22 dB for 256-QAM signals *vs.* less than 16 dB for the best 32-QAM or equivalent single-carrier systems.

First published March, 1994. Sidebar article to number 17 in the series.

Guard intervals implemented to date span from 2 μs to 32 μs. Suggestions made for future prototypes are for longer guard intervals, up to 100 μs to permit use in single frequency networks (SFNs). Net usable bit rates in the current prototypes are from approximately 25 Mb/s to 34 Mb/s in 8 MHz channels using the shorter guard intervals. 34 Mb/s matches a common telecommunications data rate in Europe. These values would be 25 per cent lower if the same parameters were used in 6 MHz channels.

Demonstrations have been conducted showing that echoes of 0 dB relative to the "main" signal can be handled by the systems under some circumstances, and increasing energy in the echoes often improves signal recovery performance. The best that can be done by single-channel systems using adaptive equalizers, currently under development in the U.S., is 3 dB for short delay echoes, increasing to 12 dB or more for longer echoes. Some combinations of echo timing do not help reception in COFDM, but rather hinder it.

The improvements in multipath performance can benefit system operation in several ways. It may allow the use of non-directional receiving antennas, or even set-mounted monopoles, instead of high-gain directional receiving antennas, but this may be problematic in lower signal level areas because of the higher signal-to-noise threshold. It can permit frequency reuse through either single frequency networks or on-channel repeaters and may permit indoor reception. It may allow the shaping of service areas and the avoidance of co-channel interference with other COFDM transmission operations despite reduced separations between transmitters.

Some aspects of COFDM systems have not yet been fully addressed by the European laboratories because of where they are in the development process. In particular, signal acquisition and synchronization systems have not been demonstrated at a practical level by any of the researchers. At least one system does not yet even have this hardware but relies on a direct connection from modulator to demodulator to supply the carrier for signal recovery. Techniques proposed range from various sync symbols sent with the data to separate pilot carriers for both local oscillator locking and clock recovery.

The Task Force identified a number of areas of concern regarding the COFDM approach. These included a higher peak-to-average power ratio than in single-carrier systems (10-13 dB for COFDM vs. 6-7 dB for single carrier systems), higher sensitivity to phase noise (requiring 20 dB better tuners than for PAL, a number that would be even higher for 6 MHz systems), possibly longer signal acquisition times (although a system potentially offering fast locking and good signal tracking was identified), and potentially higher cost for critical receiver components, some of which still require invention in order to reach required performance levels. In addition, the time required for tracking and adjusting to changes in the response of the channel make use for mobile and possibly portable reception impractical with all current proposals.

The Task Force found no benefit to cable operations from COFDM. With single-carrier systems, cable implementations might be able to double channel capacity by trading off against carrier-to-noise ratio. With COFDM so far, the trade-off of C/N is used to increase the modulation level to compensate for the lower payload data rate that results from the coding. Since the coding largely helps in multipath performance, this does not improve cable reception.

The Task Force found that there are likely to be substantial differences required between systems developed for European use in 8 MHz channels and those for North America and elsewhere using 6 MHz channels. The number of carriers used is likely to be considerably higher (one proposal is for 5,500 active carriers for 6 MHz transmission of 19 Mb/s) in order to compensate for the narrower bandwidth while maintaining the data rate and an adequate guard interval to support single frequency networks. This increase in the number of carriers would come at the price of both increased complexity and more stringent performance criteria in such areas as phase noise.

The Task Force laid out a plan for development of an appropriate COFDM scheme for utilization in 6 MHz channels. It estimated that it would take from 9 to 15 months from the beginning of a specification phase through the construction of flexible hardware combined with computer simulation to reach a state of readiness for testing comparable to the state of current QAM and VSB proposals, with the probable exception of the tuning system.

The Task Force identified two important "gating" decisions that must be reached before any effort to create a COFDM system for 6 MHz operations: (1) "that COFDM's claimed strong tolerance of multiple ghosts is a compelling advantage" and "that the somewhat lesser ability of QAM and VSB systems with multiple ghosts represents an important practical handicap in terrestrial broadcasting," and (2) that the delays that will be incurred to devise an optimized system and to invent the currently missing elements will be acceptable in the ATV process.

29

Swan Song for COFDM in U.S.?

Ever since the Grand Alliance (GA) selected the Vestigial SideBand (VSB) method as the modulation scheme for its digital television system, there has been concern expressed by some broadcasters that an opportunity might be missed for an even better transmission technology. Since the system adopted now is likely to be with us for the next 30-50 years, so the argument goes, it is imperative to explore any alternatives that might offer further improvements in signal delivery.

One such technology that on the surface offers the possibility for some improvements is Coded Orthogonal Frequency Division Multiplexing (COFDM). COFDM is receiving substantial development work in Europe and Japan. An effort was made to develop it for North America, and that effort recently underwent the scrunity of the FCC Advisory Committee on Advanced Television Service (ACATS) for the purpose of considering its certification for laboratory testing. Certification was denied.

In this installment, we will look at the development that went into COFDM for North America and see why it did not receive the certification for testing that was sought. Then we will take note of some of the possible motivations for those promoting the various transmission methods.

First published September, 1995. Number 32 in the series.

COFDM TECHNOLOGY

COFDM works by taking the data to be sent from transmitter to receiver and spreading it over a large number of carriers rather than modulating it all onto a single carrier as is done in essentially all other modulation methods. In doing so, COFDM creates a large number of parallel paths, each of which carries data at a much slower rate than that of the overall COFDM signal or of a single-carrier signal. For a complete treatment of COFDM technology, see Chapters 26 and 27.

Because of the slower data rates on the individual carriers, it is possible to intentionally ignore portions of the symbols carrying the data where echoes and multipath appear. These locations are at the beginning of each symbol period (for trailing echoes) and are where the energy from one symbol is overlapped onto the next by the delay time of the echoes in what is termed "inter-symbol interference" (ISI). By inserting a "guard interval" at the start of each symbol, it is possible to eliminate the ISI up to the length of the guard interval used.

Use of the guard interval, however, does not preclude the variations in frequency response across the channel that derive from the combination of constructive and destructive interference arriving at the receiver. Thus the levels of some carriers may be enhanced by reflected signals while the levels of others are reduced below the point at which they become unusable.

Data on the carriers reduced below the threshold is lost. To compensate for this, the data is spread redundantly over the many carriers in such a way that loss of a proportion of the carriers leads only to loss of occasional bits rather than of large blocks of data. This process, called coding, allows forward error correction (FEC) processing in the receiver to restore the data to an essentially error-free condition. The amount of coding required has a negative impact on the amount of data that can be carried in a system.

Although the spectrum of a COFDM signal consists of a large number of carriers digitally modulated, it is not generated with a bank of carrier generators and modulators. Rather, an Inverse Fast Fourier Transform (IFFT) is used to convert from a time domain signal to a frequency domain signal. Similarly at the receiver, instead of a bank of demodulators, a Fast Fourier Transform (FFT) is used to convert from the frequency domain back to the time domain.

The data on the carriers is modulated using one of the standard digital modulation schemes. Typically a form of Quadrature Amplitude Modulation is used, and any of the modulation densities, such as QPSK (which is QAM without an amplitude component), 16-QAM, 64-QAM, or 256-QAM, is theoretically possible. In order to compensate for the reduction in data throughput that results from the heavy coding required by COFDM, it generally is operated at a modulation density one step higher than would be used for conventional QAM modulation under the same circumstances. This, in turn, exacts a penalty with regard to the noise-limited reception threshold.

COFDM ADVANTAGES

Probably the most compelling argument in favor of COFDM is its potential for handling echoes and multipath signals. This becomes a particular advantage in a system where more than one transmitter is used, as in a so-called single frequency network (SFN) where multiple lower power transmitters are used to cover or extend a service area. The echo handling can theoretically extend to the level of 0 dB ghosts. If obtainable in practice, this would allow the use of non-directional (e.g. rabbit ear) receiving antennas within a dense network of transmitters.

Another advantage cited for COFDM is its ability to handle echoes and multipath without the need for an adaptive equalizer. An adaptive equalizer adds cost to a receiver, and the longer the multipath it can handle, the more expensive it becomes. A COFDM receiver requires some means for modelling the channel to compensate for differences in carrier amplitudes, but this is expected to be less costly than an adaptive equalizer.

Echoes and multipath generally are not static, and for COFDM to handle dynamic ghosting, special provisions must be made in the receiver and possibly in the signal itself. These can include the use of pilot carriers in the transmitted signal to help the receiver determine the characteristics of the channel from moment-to-moment so that it can properly set the slicing levels for data recovery from each of the carriers.

There are those who claim that handling 0 dB ghosts is only about a 3 dB improvement over what can be done with conventional modulation methods combined with adaptive equalizers. SFNs with both methods are likely to be difficult because of the costs of delivering signals to multiple sites in such a way that identical signals can be transmitted from each site – a requirement in order to treat intra-system interfering signals as echoes and multipath.

Some have suggested that the requirement for signal delivery to SFN transmitters could be handled through off-air relays on the same channel. But the problems associated with achieving sufficient isolation between transmitting and receiving antennas and gain-before-feedback to realize the necessary power levels to allow use of a relatively small number of transmitters makes this a tortuous exercise. Also of concern for SFNs in populated areas are the matters of blanketing and adjacent channel interference in close proximity to the relay transmitters.

A further advantage claimed for COFDM assumes that the rest of the world will adopt it as the standard modulation technique. It is argued that economies of scale can be obtained on a world-wide basis through use of common circuitry if North America also uses COFDM. This seems a bit specious because there is no way to know that the rest of the world will actually adopt COFDM (and there are major efforts on QAM development elsewhere in the world as well, although largely for cable applications) and because the U.S. is a big enough market that it can achieve economies of scale standing all by itself.

COFDM DISADVANTAGES

There are a number of technical challanges that must be overcome to assemble a practical COFDM transmission system. One of the most difficult is the level of phase noise that can be accumulated throughout a system. The amount of phase noise that can be tolerated drops dramatically as the number of carriers is increased.

Phase noise originates in the oscillators in a system. Thus the oscillators in the transmitter modulator and upconverter, and in the receiver tuner and intermediate frequency converters all contribute to phase noise. The tuner is the most difficult, however, because it operates over the widest range of frequencies and because it must use a phase locked loop to enable electronic control of its frequency. Achieving the required phase noise levels at consumer prices will be a major undertaking.

The peak power generated by a COFDM signal is higher than that produced by a single-carrier signal for the same average power level. The peak-to-average ratio for single-carrier signals is on the order of 6-7 dB for peaks recurring less than 0.1 per cent of the time. Recent simulation data from Europe presented at this year's Montreux symposium suggests that COFDM produces peaks more on the order of 10-12 dB, although preliminary hardware testing indicates it might only be 8-10 dB for the same time percentages.

The real question with regard to both COFDM and single-carrier signals is the average power level at which a transmitter can be operated with either of them. It is well known at this point that, with single-carrier signals, the limiting factor in transmitter average power output is the generation of intermodulation products rather than the creation of symbol errors because of peak clipping or compression. Whether COFDM operates similarly or requires more transmitter back-off to accommodate the higher peaks is still an open question – one that will haunt COFDM until it is answered.

Quickly locking the local oscillator of the receiver to the received COFDM signal is also a difficult proposition. It can be done in one of several ways – using pilot carriers if they are transmitted, looking for special code words in the data, and the like. Any method requires sampling the incoming signal for a period of time to extract the carrier information. The length of the sampling interval tends to be longer than with single carrier systems because of the slow data rate on each carrier and the consequent stretching of the time required to see the same number of samples.

COFDM PROPOSAL FOR NORTH AMERICA

The group proposing the use of COFDM for the U.S. and ultimately North American ATV system is called the COFDM Evaluation Project LLC. It comprises three commercial TV networks, three U.S. trade associations, two foreign trade associations, a foreign broadcaster, and a foreign research lab. Funding for the Evaluation Project amounts to $1.1 million.

Table 29.1 – Proposed COFDM-6 System Parameters

System Parameters[1]	System Configuration	
	Type I	Type II
Channel Bandwidth	6 MHz	6 MHz
Actual Used Bandwidth	5.6 MHz	5.6 MHz
FFT Size Required	8k	4k
OFDM Useful Symbol Duration	1101.607 µs	555.179 µs
Number of Symbols per Frame	105	105
Total Frame Duration	122.390 ms	60.927 ms
Guard Interval Period	64.0094 µs	25.0752 µs
Total Number of Carriers	6169	3109
Carrier Spacing	0.9078 kHz	1.8012 kHz
Number of Reference Pilots	49[2]	49[2]
Number of Reference Symbols per Frame	3	3
Constellation of All Carriers	64-QAM	64-QAM
Trellis Coding Rate	2/3	2/3
Payload Data Rate	≈19+ Mb/s	≈19+ Mb/s

Notes: 1. Drawn from "COFDM-6 System Technical Description"
2. Reference pilots included in documentation but eliminated
during system presentation

Formed in June, 1994, the group issued a Request For Quotation (RFQ) in June and received proposals in July, 1994. Selected as contractors were HD-DIVINE, a Scandanavian group developing COFDM for Europe, as the hardware vendor, and the Communications Research Center (CRC), a Canadian research laboratory that specializes in television among other disciplines, as the architect of the North American system parameters. The system developed for North American application is called COFDM-6 to indicate its use in 6 MHz channels.

Prototype COFDM hardware was demonstrated at the NAB convention this year and showed a remarkable amount of flexibility in the setting of the many parameters that must be optimized in such a system. Computer control allows reconfiguration of the system on command so that it is possible to balance parameter choices against performance trade-offs during system testing. Of course, once optimum values are found, the system design will be frozen and implemented in much smaller hardware.

In reality, the COFDM-6 proposal contains two systems. These are summarized in Table 29.1. The first system, Type 1 in the table, is optimized for local SFN operations. It has a long guard interval (64 µs) that is intended to enable the receiver to properly recover signals received from two synchronized stations. Approximately 6,000 carriers are required, yielding a narrow carrier spacing of about 1 kHz and a useful symbol duration of approximately 1 ms. An 8k FFT is required.

The second system, Type 2 in the table, is optimized for use in a conventional single-transmitter scheme. It has a reduced guard interval of 25 µs and uses only about half the number of carriers – a little over 3,000 – yielding a carrier spacing of about 1.8 kHz and a useful symbol duration a bit over 0.5 ms. A 3k FFT is needed for this design. Although described in the COFDM-6 documentation, this system was set aside by the proponents and not proposed for testing.

COFDM REVIEW

In response to a request from the COFDM Evaluation Project that its system be considered as a replacement for the VSB modem that is currently part of the Grand Alliance system, a Certification Expert Group was formed to consider the request. The Expert Group was chaired by Birney Dayton, president of NVision and chairman of the SS/WP-1 working party that had previously certified all of the proposed systems for earlier rounds of testing. The group met on July 20 to consider whether to certify the COFDM-6 system for testing at the Advanced Television Test Center (ATTC) as a preliminary to further comparison against 8-VSB as a possible substitute for it.

Since the ATV development process is now near an end and there is a working system undergoing field testing, the standard that had to be met by the COFDM-6 system to gain approval for testing was considerably tougher than was required of earlier proposals. Because the Grand Alliance system elements are already based upon selection of the "best of the best," the COFDM-6 proponents had to demonstrate to the expert group members a likelihood that it could exceed the performance of the current VSB modem by enough to justify a change.

In addition, Advisory Committee policy is to regard the performance of a technology as being only as good as the performance of the hardware that demonstrates it. Thus if a prototype cannot adequately demonstrate the performance of which a technology is capable, that technology will end up being surpassed by other technologies that make a better demonstration. Consequently, before approval for testing, there needed to be an expectation on the part of the expert group that in most areas the COFDM-6 prototype would achieve performance comparable to the VSB modem already tested and that in certain other areas COFDM-6 would significantly exceed VSB performance.

Following presentations on the system itself, accompanied by presentation of a study of how COFDM might be deployed to advantage in Los Angeles, several hours of questions and answers explored the potential benefits and limitations of COFDM-6 and the capabilities of the prototype. The discussion revealed a number of failings of the prototype hardware and some possible limitations of COFDM as a technology.

One of the advantages demonstrated early on by the VSB prototype is the ability to use a consumer-grade tuner. The COFDM-6 prototype currently uses an instrument-grade frequency synthesizer (costing thousands of dollars) as its local oscillator. The phase noise requirements are currently unspecified but are apparently very high. Some receiver manufacturers have estimated them to be as much as 30 dB more than current consumer electronics practice, requiring invention of new technology to make a consumer tuner for COFDM-6 practicable. Others think the COFDM phase noise penalty might be only on the order of 10 dB above the requirements for VSB. How these relate and what is really required, aside from "considerably more," is unclear at the moment.

Similarly, the COFDM-6 prototype uses a 12-bit analog-to-digital (A/D) converter that operates at a sampling frequency in the 20 MHz range. Such a component costs several thousand dollars today. Concern was expressed over how quickly such a device could be brought down to consumer prices. The proponents pointed to data indicating that a smaller number of bits might actually be required for real world hardware, but there was never any conclusion to the discussion that would have eliminated the concern.

The test system uses data frame preambles to acquire carrier lock. The test system sends preambles every 122 ms and must sample 25 preambles before locking. The test system cannot utilize every preamble because of software processing time and consequently takes about 8 seconds to acquire carrier lock. This might be reduced to 4 seconds with a modification, and it was estimated that a production modem could acquire in less than 1 second. Given consumers' penchant for quick channel changing, this could prove to be still too long since the rest of the decoder must yet lock up after the modem does. The VSB modem locks in a couple hundred milliseconds or so.

Dynamic ghosts are a fact of propagation life, as was thoroughly demonstrated in the Charlotte field tests of the VSB modem. The COFDM-6 prototype was not designed to handle dynamic ghosts, so this capability of COFDM could not be fully tested without significant modification to the hardware. The VSB modem has been demonstrated to be quite adept at handling the dynamic ghosts found in field testing.

In most, if not all, of the digital modulators, a Surface Acoustic Wave (SAW) filter is used to shape the signal so that it occupies the channel as fully as possible without spilling into the next channel. The spectrum plot of the COFDM-6 prototype showed an aberration in the SAW filter response that would have allowed undesired energy to fall into the upper adjacent channel. This was certainly correctable through substitution of a properly designed SAW filter, but it was noted as an item that would delay the availability of the system for testing if the real characteristics of COFDM-6 interference, particularly into NTSC signals, were to be determined.

The conclusion to the discussion was a list of five issues identified as necessary to be addressed by the COFDM Evaluation Project before ATTC testing could be conducted to permit a "bake-off" level comparison between COFDM-6 and 8-VSB. The expert group members were polled to confirm a consensus that the current COFDM-6 system is not ready to test and that there was not a demonstration of superiority of COFDM.

The COFDM proponents in attendance were asked how long it would take to resolve the various issues. After caucusing, they indicated that they could return the following week with responses to the issues raised, indications of the improvements to be implemented, and a schedule for incorporating the changes. A further poll of the expert group members indicated a unanimous disbelief that COFDM could be proved likely to be superior to VSB in that time.

Authorization for the July 20 meeting had been given by Advisory Committee chairman Dick Wiley on the basis that a decision on testing of COFDM-6 would be made by its conclusion. Given the likelihood that a follow-up meeting a week later would still not achieve that result, chairman Wiley's representative present at the meeting indicated that he could not recommend authorization of a further meeting. And that ended the matter.

The shame of this result is that most of the issues raised at the meeting were identified in late 1993 by the Advisory Committee's Transmission Expert Group during a visit to the European laboratories that have been developing COFDM. They were, in fact, spelled out in the Expert Group's report and distilled in a sidebar by your intrepid writer, [included herein as Chapter 28]. Rereading that report today is hauntingly like reading the report just above of the July 20 meeting. One has to assume that the reason more was not done to address these known problems is that COFDM is just in too early a stage of development for the prototype's builders to have dealt with them while also coming up with solutions peculiar to the needs of 6 MHz operation.

MODULATION MOTIVATIONS

There are a number of reasons why various parties might prefer different modulation methods. Some are technical and some are techno-political. There are those who see the interest in COFDM by broadcasters as being an effort to stall the process and thereby avoid having to implement ATV so soon. In reality, broadcasters have, or should have, a number of concerns about how to reach the audiences they reach today, and some see COFDM as a potential answer to those concerns.

When you move out of the major markets and away from the coasts, especially west of the Mississippi, there are large expanses that are served by just a few television stations. The combination of extended VHF propagation and translator stations helps to provide television service to many who would not otherwise receive it. In order to receive "local" programming, people still watch signals that are far below Grade B and full of multipath.

As ATV comes along, some translators will be bumped from their current channels, and the signal fades that now just cause noisy pictures at long distances may result in total loss of service because of the "cliff effect" of digital signals. It will also become increasingly difficult to find channels for translators as the industry is compacted into less spectrum with twice as many transmitters. The smaller coverage areas that can be achieved with reasonable amounts of power on UHF as compared with (especially low-band) VHF only exacerbate these issues.

Broadcasters who are thinking ahead to future realities are consequently concerned about how to implement translators and "gap fillers." They see the claims for COFDM as potentially addressing their concerns, so it is attractive. For the reasons outlined above, they probably also need to be thinking about how to secure the same objectives with the existing modulation method – VSB.

Among equipment manufacturers in North America, all but Zenith are pushing QAM outside the broadcast environment. For cable, for wireless cable, and for the hybrid fiber/coax networks of the telcos, 64-QAM and possibly 256-QAM are the modulation methods of choice. Set top converter/decoders are being designed and demonstrated, and orders for them are being taken. COFDM offers no benefits in wired environments, and VSB comes with a patent royalty obligation to Zenith, while QAM is in the public domain. Even in the rest of the world, QAM has many adherents and is seeing tremendous development effort, albeit largely for cable applications.

Zenith, of course, is the originator of VSB and has the most to lose if something else is adopted for the U.S. ATV standard. It is only VSB's adoption in the Grand Alliance system that has brought interest from other manufacturers. Some think that VSB was selected by the GA because no other major Zenith contributions remained in the system, but in fact, it won the bakeoff, if only by a small margin. In other areas than broadcast, Zenith (and perhaps Goldstar, now commonly owned with Zenith) is the only flag bearer for VSB.

Supplanting VSB in the GA system now would be an extreme blow to Zenith, which in some ways has "bet the business" on its participation in ATV. It might even result in blowing up the GA itself. It would behoove the GA to directly address broadcasters' concerns about how to extend coverage and fill in gaps using VSB, as it has said can be done, if it wants to avoid having COFDM reintroduced as an alternative after the Advisory Committee has run its course.

Naturally, if the FCC's reopening of many of the already decided questions causes a long delay in adopting the system, technology will advance, and many things once again may be up for grabs.

PART 9

Standard Definition Television

Standard Definition Television (SDTV) seems like such a simple thing. After all, it is what we have been doing all along, isn't it? Well, not quite. In fact, the definition of SDTV and the choices of what formats to allow to be used as SDTV signals were the subjects of considerable debate in the FCC Advisory Committee and the Advanced Television Systems Committee. Many issues were raised that had more to do with the politics of the convergence of the broadcasting, consumer electronics, and computer industries than with anything purely technical. In the end, a compromise was reached, but still not everyone was happy with the outcome.

Distributors of programming outside the broadcast realm, such as cable operators, wireless cable operators, direct broadcast satellite operations, telephone companies, and the manufacturers of set top decoders that supply them generally have not had to deal with these issues on a large scale. Because they are not subject to the setting of standards through a regulatory process, they can pretty much decide between themselves what they want to do and then proceed with it. Thus the big organizations decide what they want, place big orders, and the smaller organizations ordinarily go along. In general, SDTV (if the term is used at all) has meant 525-line, 59.94 Hz, interlaced, component signals, often decoded from NTSC, with provision sometimes made for 16x9 aspect ratio signals in addition to the standard 4x3.

In Chapter 30, the flavor of some of the wrangling within the broadcasting standard-setting process is captured. Also in that chapter, somewhat similar outstanding (when written) audio issues are addressed. (Nobody ever said the articles could only deal with one subject at a time.) In Chapter 31, we come back to the SDTV issue after the decisions have been reached and give the final outcome. Just as important, in Chapter 31, we look at the issues of compatibility between HDTV and SDTV within the ATV transmission scheme and examine a method being developed to allow display of HDTV images on SDTV displays, using SDTV-technology and SDTV-cost decoders.

30

The Meaning of SDTV, plus Audio Issues

You might think that, after all the years of 525-line television, the meaning of the term Standard Definition Television (SDTV) would be readily apparent. In the world of Advanced Television, at least, it's not. You might also think that defining the number of audio channels for a new television system would be a simple matter of picking a number — 5 or 5.1 perhaps. Again, in the world of Advanced Television, it's not.

These two issues are at the center of a great deal of current debate and disagreement. The decisions to be made about them shortly will affect the future structure of the television experience that will be delivered to viewers and may also impact the structure of the businesses that produce and deliver the programs that constitute that experience.

We will spend our time together in this chapter looking at the various aspects of the separate but similar debates. Then in the near future, when we read about the decisions that have been made, we'll have some understanding of the background to the announcements.

SDTV DEFINED

SDTV is one of two terms applied from the point of view of Advanced Television to the television systems currently in use. The other term is Conventional Definition Television (CDTV). CDTV applies to analog encoded signals such as NTSC and PAL and presumably to anything else that could be

---

First published June, 1995. Number 29 in the series.

called "conventional television." Because of the very specific definitions of these signals and the relationships between their subcarriers and their interlaced line structures, there is little room for debate over what they are, although there could be some minor disagreement over just which ones to include in the category.

SDTV, on the other hand, is a term much more open to interpretation. As defined in the recently approved ATSC "Digital Television Standard for HDTV Transmission," it means "a *digital* television system in which the quality is approximately equivalent to that of NTSC. This equivalent quality may be achieved from pictures sourced at the 4:2:2 level of ITU-R Recommendation 601 and subjected to processing as part of the bit rate compression. The results should be such that when judged across a representative sample of program material, subjective equivalence with NTSC is achieved."

All of these words mean that SDTV can be anything digital that someone can interpret as being roughly equivalent in subjective quality to NTSC. There is no definition of equivalence, there is no guidance on the qualities of NTSC to which an SDTV system is to be equivalent, and there is little guidance on how to judge the equivalence. This leaves room for each proponent of a particular approach to SDTV to interpret the specific system proposed as fitting the definition. Indeed they all may.

With this wide latitude in what constitutes a Standard Definition system or signal, it is little wonder that there are many different suggestions being considered in the standards setting process for ATV. Clearly, all possibilities cannot be accommodated without creating a very complex and expensive system, particularly in regard to the decoder, of which there will be millions. Thus it is necessary to narrow the possibilities to a few that can be economically implemented in all receivers. It is the selection of the best alternatives that is now the focus of the ATV standards setting organizations.

ATV SDTV STANDARDS

There are two organizations principally involved in determining the SDTV standards that the industry will use and that will be submitted to the FCC as part of the recommendation for an ATV system. These are the Advanced Television Systems Committee (ATSC) and the FCC Advisory Committee on Advanced Television Service (ACATS).

The Advisory Committee has responsibility for making the actual recommendation to the Commission. It has served as the policy making body with regard to the ATV system and its operation. The ATSC was assigned the task of writing and adopting the actual standards that will document the ATV system. The FCC Rules are then expected to reference the ATSC standards rather than directly incorporating all the minutiae of the system details. Since the ATSC's job in this instance is that of documentor, when it runs into something requiring further definition as it has with regard to our subject matter this time, it bounces it back to the ACATS for decision. That is where things stand as this is written. Perhaps by the time you read it, decisions will already have been made.

For a long time, the subject of carrying other-than-HDTV programming as part of the ATV signal was *verboten* within the politically correct Advisory Committee process. All of this began to change just about a year ago, and efforts have been under way for roughly the last six months to rapidly decide on SDTV standards to go along with those for HDTV so that both can be submitted to the FCC together as part of the recommendation. This has put a great deal of pressure on completing in short order the decision making on some very difficult issues.

SDTV ISSUES

The arguments and indeed the parties to those arguments with respect to SDTV standards are essentially the same as those that existed with respect to the HDTV standards. Basically, the issues are interlaced vs. progressive scanning, the use of a square sampling matrix vs. non-square pixel aspect ratios, and the fundamental horizontal and vertical resolutions to carry.

Also thrown into the mix in the case of SDTV is whether to limit the signals to the Main Profile at Main Level (MP@ML) of MPEG-2 or alternatively to allow them to extend into the Main Profile at High-1440 Level (MP@H14L). (See Chapter 5 for a detailed look at profiles and levels.) Limiting SDTV to MP@ML has the effect of eliminating some of the choices for signal characteristics that are very important to certain interests. We'll see the results of this shortly.

As in the debates over which raster formats to use for HDTV, the positions taken by the various participants in the discussion on SDTV depend primarily upon the industry backgrounds from which they approach the problem but with some very interesting cross-overs from earlier discussions. Oversimplifying somewhat, those interested principally in compatibility with computers favor progressive scan and square sampling matrices while those coming essentially from a television production, distribution, or manufacturing background tend to prefer interlaced scanning and non-square pixels for best compatibility with existing 525-line signals and equipment.

Without naming names, two of the four networks participating in the discussions now seem to favor the progressive scan/square pixel approach, which is one more than heretofore. Similarly, there are now several TV set manufacturers that appear to prefer progressive scan and square pixels, at least optionally if not exclusively. There do not seem to be any members of the computer community who have moved into the interlaced/non-square pixel camp, however.

It should be noted that the current discussion is over the format(s) to be used for **transmission**, not production. Because of the digital nature of the source signals and of the compression process, it is possible to use almost any production input, although it may not be easy to obtain a high quality conversion from one format to another.

It should also be recognized that virtually all television sets and virtually all computers set up to handle off-air television signals will require the capability to receive NTSC signals for a very long time to come. If they incorporate progressively scanned displays, they will require a de-interlacer to permit presentation of the interlaced NTSC signals. Because TV sets (and consumer computers) must be seriously cost-reduced to make them competitive in the consumer marketplace, the de-interlacers built into them likely will not be as high in performance as those that could be afforded in broadcast plants.

ARGUMENTS

The debate over interlaced versus progressive scanning continues to be somewhat of an argument over religion. Nearly everyone involved admits that, theoretically, progressive scanning should yield better pictures. Unlike the situation with the HDTV battles, however, in which there were limits claimed in what production equipment could be economically produced in the transition period, this time around the arguments against use of pro-scan have to do with the additional data bandwidth required for pro-scan versus interlaced rendition of the same images and the fact that MP@ML does not allow for progressive scanning above a 30 Hz frame rate.

It has been estimated by the proponents of progressive scanning that there will be about a 10 per cent penalty in additional data bandwidth required to transmit a progressively scanned image obtained by upconversion from an interlaced source over directly sending an interlaced signal. Why pay the penalty, both in equipment cost and bits used to carry the signal? The argument is that a de-interlacer at the studio, which serves thousands of receivers, can be much more nearly perfect in its results because a much more complex and expensive one can be afforded than can be put into every television receiver. This, of course, assumes that the majority of receivers will require de-interlacing or some other form of line multiplication for display that can be best accomplished or at least begun at the transmitter.

Thus sets with built-in de-interlacers might only use them with off-air/off-cable analog NTSC signals, not with those arriving digitally compressed. Those sets that had interlaced displays (including set top converters feeding NTSC receivers) would interlace (or re-interlace) the images through application of a simple vertical filtering and line decimation process that should be easy to implement with high quality in receivers. All signals transmitted could be in progressive form, with those that originated in interlaced form being de-interlaced before transmission.

There is a further penalty in bit rate for transmitting a progressive source versus sending the same image in interlaced form. The bit rate required is estimated to be 30-to-50 per cent higher than needed for just transmitting an interlaced source in its native form. This is a large enough increase that it will certainly result in reducing the number of programs that can be carried in the name of improved image quality. Because of the price to be paid in reduced capacity, there are some who favor permitting both interlaced and progressive forms of transmission to be available, depending upon whether a particular situation calls for more capacity or higher quality.

MP@ML DEBATE

Another part of the discussion swirls around the question of whether to limit SDTV to fitting within the Main Profile at Main Level of MPEG-2. This will be of concern if there are receivers or set top converters built that only have capability for receiving SDTV signals and not HDTV signals. If they have capacity for receiving HDTV signals, by definition they have the capacity to handle higher levels than MP@ML. To meet the requirements of MPEG-2 conformance, they must accept everything at a lower level within the same profile, thus including the MP@H14L required for 59.94/60 Hz progressively scanned SDTV.

Interestingly, a number of receiver manufacturers have come forward to indicate that it should be possible to build relatively inexpensive receivers or set top boxes that can interpret all the ATV formats including HDTV signals for presentation on 525-line interlaced displays. There have also been indications from several manufacturers that a small, inexpensive modification to the chips needed for MP@ML would allow them to handle a limited range of signals from MP@H14L sufficient to cover the values proposed for progressively scanned SDTV. An issue with these approaches would be that, for every new format that is allowed, receivers would have to provide filters to convert that format to the display's format.

In the end, the issue over MP@ML seems to boil down to whether to make broadcast ATV SDTV signals compatible with the set top boxes currently in the design process or those few MPEG-2 compatible units that have already been installed, neither of which incorporates the proposed modifications. If the assertions concerning the ability to build inexpensive receivers that can make non-HDTV pictures from HDTV signals and about the ease of modifying MP@ML chips to handle low-MP@H14L both prove to be true, this issue would seem otherwise to become moot.

ASPECT RATIO AND RESOLUTION

There seems to be general agreement among participants in the discussions that both 4:3 and 16:9 aspect ratios must be accommodated in their native formats. This means that 4:3 images will be carried efficiently in that mode rather than being arbitrarily positioned within a 16:9 raster and then extracted using pan and scan data as had been proposed by some. Pan and scan data will be carried, if provided with the source signal, to allow extraction of 4:3 images from widescreen sources according to control data inserted at the source or elsewhere upstream.

Far more contentious is the matter of the resolutions to be allowed, or more correctly the numbers of active pixels per line and the numbers of active lines per picture to be used. A wide variety of suggestions have been put forward. These include values from the 352x240 (HxV) of MPEG-1 (also called Source Input Format – SIF) to 480x360 (progressive only) to 720x480 up to 848x480 (progressive only, in 16:9 aspect ratio, with nearly square pixel matrix). Many of these are familiar values and the real questions surrounding them are just which ones will be most useful and should therefore be included in the allowed matrix of transmission signal formats.

The suggestion of 360-line progressive formats is a relatively new one including 480x360 for 4:3 aspect ratio and 640x360 for 16:9. This scheme provides a square pixel matrix in either aspect ratio. Its most interesting characteristic is that, because it is progressive, it has a higher Kell factor than interlaced raster formats($\approx$0.9 vs. $\approx$0.7 for interlaced signals) and results in about the same (static) vertical resolution as 480-line interlaced signals. Of course with dynamic images, interlaced signals can fall to half the vertical resolution while progressive signals retain full vertical resolution. The 360-line approach would also result in about the same number of bits after encoding as would a 480-line interlaced signal. One difficulty with it is that, because it has a 59.94 or 60 Hz frame rate, it falls outside MP@ML.

It is very difficult to predict where all of this will come out. There are three possibilities: (1) the MP@ML camp will win, and there will be no 60 Hz progressive transmission of SDTV (although 24 and 30 Hz frame rate progressive will be allowed); (2) the progressive side will triumph, and there will be no interlaced transmission; or (3) some compromise will be struck allowing both forms, with the user able to choose what is best for each application.[15] Lest you think that choice 3 is a panacea, keep in mind that the more possibilities there are, the more difficult it will be to smoothly concatenate program segments for program continuity integration purposes such as commercial insertions. (For more on this matter, see my paper "Switching Facilities in MPEG-2: Necessary But Not Sufficient" in the Proceedings of the 1995 SMPTE Advanced Television and Electronic Imaging Conference and in the December, 1995, issue of the *SMPTE Journal*.)

---

[15] The SDTV formats to be included in the ATSC standard and in the ACATS recommendation to the FCC were ultimately decided. See Chapter 31 for a full discussion of the decision and a table showing the final outcome.

AUDIO MATTERS

A similar conundrum faces the audio portion of the standards for both HDTV and SDTV. The question at issue arises from a deeper understanding of the implications of the provisions for several different audio service types in the ATSC standard. There are two types of main audio services and six types of associated audio services defined.

The two main audio services are called Complete Main (CM) and Music and Effects (ME). The associated services are for the visually impaired (VI), for the hearing impaired (HI), a dialogue channel (D), a commentary channel (C), an emergency channel (E), and a voice-over channel (VO). A complete audio service consists of a main audio service or a main audio service combined with an associated audio service. The receiver is required to have the capability to simultaneously decode one main service and at least one associated service.

The idea of these services and combinations is to allow different complete services to be assembled from components that are shared between alternative services without having to be duplicated for each of them. Thus it should be possible to send one ME main service and add to it (mix with it) one dialogue channel to make a complete service. If several different dialogue channels are sent in different languages, for example, different complete services can be delivered for each language without using the data bandwidth that would be required to code each of them as a Complete Main (in essence, simulcasting them).

The difficulty hinges on the requirement to simultaneously decode a main service and an associated service. Since simple receivers and set top converters are likely to be able to interpret HDTV signals as well as SDTV signals, as discussed above, it becomes necessary to use the same audio system for both HDTV and SDTV. Since it is intended to use the same audio system for both, simple receivers would be obligated with the same requirements as are large ones. This could be a burden in a small, hand held receiver or in a set top converter. Yet without the same basic capabilities in all receivers, many of the features provided become meaningless for broadcasters.

An example should prove the significance of the last statement. If a broadcaster sends an ME main plus a dialogue channel and there are any receivers that do not have the capability to decode and mix the dialogue channel, then the output of those receivers will have no dialogue. This would mean effectively losing part of the audience, an unacceptable consequence for most broadcasters.

Another implication is that, in order to send multiple languages, a broadcaster is forced either to send an ME channel as the main audio service with a dialogue channel for the primary language or to send both a complete main for the primary language and an equivalent bandwidth ME channel to support additional languages. If it is desired to be able to position the dialogue other than in the center front speaker, the latter, redundant method becomes mandatory.

There are several possible alternatives under consideration to handle the various cases. Matters only become more complicated when the uses of the other associated services are explored. We won't go into them here. It is important to recognize, however, the fundamental conflict between the interests of broadcasters and of receiver manufacturers on this issue. The broadcasters need receivers to be capable of supporting the most efficient use of the data stream by providing multiple decoders with the necessary mixing capability so that duplication of material does not waste bits. Receiver manufacturers, on the other hand, need to minimize the cost and size of receivers in order to be competitive.

It will take careful understanding of each other's needs, good will, and likely some compromise to come to a result that will best serve the needs of the viewing public and of the various stakeholders in this issue. Time is running short to meet the schedule for submission of a recommendation to the Commission. There is more hard work ahead.[16]

---

[16] In the end, it was decided to require only a single decoder in receivers conforming to the ATSC standard. Broadcasters must transmit at least a Complete Main signal; associated services are permitted to be transmitted on an optional basis. This is not the most efficient use of bits, but it does provide for manufacture of small and/or inexpensive receivers to present digital broadcast transmissions. It also allows the development of additional services for which consumers can be convinced to purchase receivers with the extra features to support them.

31

Extracting SDTV Images from HDTV Streams

One of the difficult questions still to be answered regarding broadcast advanced television is the nature of the relationship between high definition television (HDTV) and standard definition television (SDTV). This relationship has many aspects including technical, regulatory, production, and consumer acceptance matters. We looked at the problems of defining SDTV in the last chapter. This time around, we'll catch up a bit on developments in the technical definition of SDTV and quickly move into the other areas that are likely to be more important in their influence on the success or failure of broadcast ATV.

SDTV DEFINED, REPRISE

When we tuned in last time to the ongoing saga of the definition of SDTV, our players were rangling over the numbers of pixels to include in the active picture area, both horizontally and vertically, the frame rates to allow, whether to insist on exclusive use of progressive scan, and the aspect ratios to be allowed for the various combinations. Also mixed into the argument was the question of whether to restrict SDTV signals to the Main Profile at Main Level (MP@ML) of MPEG-2, thereby precluding 60 Hz progressive scan from the table of permitted combinations of SDTV operating values.

At that time, the table included active picture areas of 352x240 (HxV) pixels, 480x360, 640x360, 640x480, and 704x480, with the first of the 360- and 480-line systems having a 4:3 aspect ratio and the second ones having

First published November, 1995. Number 34 in the series.

Table 31.1 – "All-Inclusive" SDTV Transmission Formats List
w/MP@ML Restriction

Format Designation	Active Lines	Horizontal Pixels	Aspect Ratio	Picture Rates	Square Pixels
CCIR Rec. 601	480	704	4:3/16:9	60I 30P 24P	No
VGA	480	640	4:3	60I 30P 24P	Yes
360	360	480 ——— 640	4:3 ——— 16:9	60P 30P 24P	Yes
MPEG-1	240	320	4:3	30P 24P	Yes

16:9. This is shown in Table 31.1, which includes only the SDTV values that were discussed. Allowed frame rates included 24 Hz and 30 Hz progressively scanned and 60 Hz with interlaced scanning. (The integer frame rates given are understood to include as alternatives the corresponding NTSC-related frame rates obtained by multiplying the included frame rates by 1/1.001.)

The table included the 352x240 pixel matrix for compatibility with MPEG-1. 640x480 was intended to be compatible with VGA computer monitors. And the 704x480 raster was to be compatible with CCIR-601/SMPTE 125M systems.

The 360-line entries, which were to be exclusively progressively scanned, were included to provide essentially the same resolution as would be obtained from 480 lines of interlaced scanning, based on their different Kell factors. They were expected to provide approximately the same image performance with the same bit rates as would 480-line interlaced signals, although nobody had built and demonstrated any equipment using 360 lines – at least not as part of the ATSC or FCC Advisory Committee efforts. The 360-line systems were touted as providing the motion rendition advantages of progressive scanning and the square pixel matrix that are near and dear to the hearts of some in the computer industry and elsewhere.

With time, it became clear that the number of entries in the table was too large. Since every receiver conforming to the standard would be expected to accept all of the table entries as inputs, such receivers would be inordinately complex. This was especially important considering that one of the goals of

Table 31.2 – Final HDTV and SDTV Transmission Formats
from ATSC Standard

Format Designation	Active Lines	Horizontal Pixels	Aspect Ratio	Picture Rates	Square Pixels
1080	1080	1920	16:9	60I 30P 24P	Yes
720	720	1280	16:9	60P 30P 24P	Yes
CCIR Rec. 601	480	704	4:3/16:9	60P 60I 30P 24P	No
VGA	480	640	4:3	60P 60I 30P 24P	Yes

SDTV was to permit low cost receivers to be built, in addition to providing a mechanism for transmitting a multiplicity of programs over the same channel.

A meeting was held by the FCC Advisory Committee expert group responsible for video matters at which the goal was to resolve the unhappiness that existed over the previous table. After considerable compromise from most parties represented, the SDTV entries were reduced to just the two 480-line arrangements. These were then added to the table in the ATSC standard document that shows the permissible transmission formats, both HDTV and SDTV, which is reproduced here as Table 31.2.

OF CHICKENS AND EGGS

An important statement two paragraphs back is that every receiver conforming to the standard would be expected to accept all of the table entries as inputs. This is necessary because a receiver would display a blank screen (or worse if it weren't sufficiently intelligent) if it received a signal in a raster format that it was not designed to handle. If a large number of receivers were sold that could not receive a particular raster format, that format would become unusable for broadcasters because they would cut off parts of their audiences through its use.

So what about the same consideration with regard to the various SDTV and HDTV raster formats? And what happens when NTSC transmissions are turned off? The same arguments can be made as in the case of SDTV alone. For the HDTV raster formats, it is understood that receivers must accept either of them, making any necessary conversions to the particular display format used. But currently there is no such requirement for receivers regarding inclusion of both HDTV and SDTV raster formats.

One particularly troubling scenario becomes possible when there is no linkage between HDTV and SDTV in receiver capabilities. If SDTV receivers are the first sold in large quantities and broadcasters transmit SDTV programming to serve them, any HDTV programs then transmitted will not reach the audience viewing the SDTV receivers. The HDTV programs will reach only those viewers with HDTV receivers – a smaller audience, thereby increasing the production and distribution cost per viewer of programming that will already be more expensive than with standard techniques. The converse is not true, by the way, because HDTV receivers, by their natures, are capable of receiving SDTV transmissions, as we will see shortly when we look at the designs of receivers.

When NSTC transmission is shut down, in a decade or so, some means will be needed to keep all of the television receivers from having to be thrown out. Discussions to date have centered on a set top converter that would receive digital transmissions and output NTSC signals for use on older TV sets. The cost of such converters is, naturally, a major consideration, and we will discuss it momentarily. Assuming for the moment that SDTV-capable set top converters will be substantially less expensive than HDTV-capable versions, one can readily see that many more SDTV set top converters would be sold than HDTV units. This would tend to bias the audience away from HDTV, again narrowing the audience for HDTV programs.

Since receiver uptake by consumers and program availability to consumers go hand-in-hand in chicken-and-egg fashion, as was dramatically demonstrated by the correlated sudden shifts in color television programming and color set penetration in 1965, any market tendencies that reduce access to HDTV programming may well stifle or at least delay its implementation.

Because of these considerations, the FCC has included in its latest Notice of Inquiry (NOI — with comments currently due on November 15 [,1995,] and reply comments due next January 12 [,1996,]) some questions about what it should do in regard to the possibility of mandating various kinds of capabilities in receivers. For instance, during the transition period, should it permit the manufacture and sale of receivers that receive only NTSC, SDTV, or HDTV signals? Should it permit receivers that handle any two of the signal types? Or should the Commission mandate that all receivers must be capable of dealing with all three? Should it require that HDTV signals must be displayed as true HDTV pictures, or would display as lower resolution SDTV images be acceptable? The answers to these questions, through their impact on the workings of the marketplace, may well determine what the "face" of broadcasting looks like in the future.

DECODER AND CONVERTER COSTS

Any examination of the possibilities for handling SDTV and HDTV signals must include the cost of decoders as one of its elements. ATV decoders are MPEG-2 decoders, and their costs comprise, in roughly decreasing magnitude order , picture storage memory, channel buffer memory, the Inverse Discrete Cosine Transform (IDCT), the Inverse Quantizer, motion vector processing, a run length processor, a variable length decoder, some post processing, a syntax parser, a memory controller, and a digital-to-analog converter. These functions are all contained on several integrated circuits.

The relative costs of the components for HDTV and SDTV decoders can be determined by summing the costs of the individual functions. While it is beyond our scope here to do such an analysis, it will be instructive to look at just a few of the major functions.

By far, the cost of memory predominates in the cost MPEG-2 decoders, at least into the foreseeable future. A practical HDTV decoder will require some 12-16 MBytes of DRAM, with a current cost in the range of $300-$400, and the likelihood that it will cost more than $100 through much of the transition period. By contrast, a cost-effective SDTV (MP@ML) decoder can be built using 2 MBytes in a single 16 Mbit chip, at a cost currently well below $100 and likely to fall to a fraction of that amount over the transition period.

Another important factor in the cost of the devices in a decoder is the speeds at which they must operate. For example, the channel buffer for an HDTV decoder must operate at peak data rates up to 140 MBytes/second, while the channel buffer for an SDTV decoder peaks at 23 MBytes/second. This difference can have a significant impact on the number of parallel paths required within a device or the number of devices required within a circuit. Either way, cost is directly impacted.

Component costs do not directly indicate the prices at which equipment will sell; there are assembly and distribution costs, as well as overhead and profits, that must be added. But they can provide a good indication of the relative prices to be expected. If we consider that digital set top converters based on MPEG-2 MP@ML are currently being offered to cable and wireless cable operators and to the telcos in the range of $300-$400 (which can be considered wholesale prices and which might drop perhaps under $200 in a decade), we can see that equivalent HDTV set top converters would have to sell for several times as much. Some estimates have put the cost of silicon components for an HDTV decoder at 6-10 times the cost of the equivalent components for an MP@ML decoder. The other components are not any more expensive, so the selling prices of HDTV relative to SDTV decoders are likely to fall between the two ratios, perhaps 4-6 times as much.

When considering the cost of a digital set top converter, as will be used to prolong the utility of NTSC receivers beyond the period of NTSC transmissions probably more than a decade from now, it is important to recognize that the price relative to the price of the television set with which it will be used is also significant. Similarly, the cost of a decoder relative to the cost of the remainder of a television set into which it is built is also important. If the price is so high that it makes the continued use of existing sets or the purchase of new sets too expensive, people will either reduce their purchases of television equipment or turn to delivery media that do not require the more expensive equipment. Also, those who can afford the cost may make the necessary purchases while others may not be able to do so. This has significant social and public policy implications.

DECODER PERFORMANCE REQUIREMENTS

In contemplating the types of decoders to be used with ATV signals, it is helpful to keep in mind the impact of display size on what is perceived by the human visual system. A small display cannot exhibit a high definition image. A moderate size display can show an HDTV image upon close viewing, but it won't be perceived by the viewer at normal television viewing distances. A large screen display can present a high definition image at normal viewing distances, and it will be appreciated by the normal viewer.

Many studies have shown that, in North America at least, the average viewer tends to favor image brightness over resolution if the two cannot be had together. Since there is a trade-off between screen size, brightness, resolution, and cost, it is likely that the earliest large screen television receivers capable of decoding HDTV will not be able to display it at full resolution. They will have reduced resolution in order to maintain acceptable brightness at a manageable cost. As some of the newer display technologies soon to emerge from the laboratory achieve volume production, this may change, but only after a number of years accumulated experience moving them along the learning curve.

All of these factors suggest that, if it were possible, the optimum situation would be to use decoders that could accept as inputs any of the raster formats included in the table, from SDTV through HDTV, and whose resolution performance was matched to the displays with which they were associated. Assuming that lower performance decoders would be lower in cost and that higher performance decoders would be higher in cost, they would also be matched with displays at corresponding points in the display cost continuum. Thus only a high priced, large screen display would require a high performance decoder, and only it would be burdened with a high cost decoder. Smaller displays would still be able to exhibit HDTV programming, but they would only have to bear the cost of correspondingly lower cost decoders. This would be rather like having one's cake and eating it, too.

ALL-FORMAT ATV DECODERS

It may not be so far-fetched to think that the optimum situation we just described can be approached, if not fully achieved. A number of companies are

Table 31.3 – Complexity Comparison of All-Format ATV Decoder

Function	ATSC Standard HDTV	All-Format ATV Decoder	MP@ML (SDTV)
Pre-parser	—	<10,000 gates 19 Mbits/sec (partial decode)	—
Chnl Buffer Size Bandwidth	8 Mbits 140 MBytes/sec	1.8-4.3 Mbits 23 MBytes/sec	1.8 Mbits 23 MBytes/sec
Total Off-Chip Memory Requirements	96 Mbits specialty DRAM	16 Mbits DRAM	16 Mbits DRAM
Syntax parser/ VLD	93 M coefficients/sec	15.5 M coefficients/sec	15.5 M coefficients/sec
IDCT	1.5 M blocks/sec	1.5 M half complexity simple blocks/sec (HD) + 240 K full 8x8 blocks/sec (SD)	240 K full 8x8 blocks/sec (SD)
Upsample/ Downsample	—	1000-2000 gates?	—
Decodes HDTV	Yes	Yes	No
Decodes SDTV	Yes	Yes	Yes

known to be working on what the FCC in its NOI has called All-Format decoders. The objective of these efforts is to develop a decoder with the complexity and cost of one in the MP@ML that produces SDTV outputs from signals with transmission raster formats anywhere in the table. The reason for concentration on the low end of the decoder scale is that HDTV (high level) decoders, in order to achieve MPEG-2 compliance, are required to decode the signals at lower levels within their (main) profile. Since they include more than enough memory and processing capacity, this will be easy. Adding NTSC reception to the combination will be relatively low in cost given everything else that must already be included.

Possibly the most advanced in the work they have been doing in the area of all-format decoders has been the Hitachi America Advanced Television Systems Laboratory. This group conducted public demonstrations of their progress to date at the recent SMPTE conference in New Orleans. The

Table 31.4 – Memory Usage of Reduced Resolution Decoding in All-Format Decoder

Active Horiz. Pixels	Active Vert. Pixels	H Scaling Factor	V Scaling Factor	Down-Sampled Horiz. Pixels	Down-Sampled Vert. Pixels	Number of Frames Stored	Down-Sampled Memory Required	Free Memory w/16 Mbits DRAM
1920	1080	3	2	640	540	3	12,441,600	4,335,626
1280	720	2	2	640	360	2	5,529,600	11,247,616
720	480	—	—	—	—	3	12,441,600	4,335,616

Note: Last row shows SDTV format for comparison

description of the operation of an all-format decoder to follow is based on the Hitachi work. It is acknowledged that other organizations may have different methods for accomplishing the same results.

The principal objectives in an all-format decoder are to produce a high quality SDTV output from both SDTV and HDTV inputs while using an MP@ML decoder with as little modification as possible. This is accomplished by adding to the MP@ML decoder small amounts of circuitry that allow it to downconvert HDTV signals while reducing the memory required for HDTV decoding, reducing the processing speeds, and reducing the processing complexity involved relative to what is needed in an MP@HL (HDTV) decoder. This added circuitry is estimated by Hitachi to increment the cost of the LSI ICs and RAM for an MP@ML decoder by about 10 per cent over the cost for the same devices in an ordinary SDTV decoder, compared to an estimated six- to ten-fold multiple for the LSI ICs and RAM for a true HDTV (MP@HL) decoder.

To help in understanding the Hitachi design, Figure 31.1 is the block diagram of an MPEG-2 MP@ML decoder with the the added circuitry shaded. Table 31.3 is a complexity comparison between an ATSC standard HDTV decoder, the all-format ATV decoder, and an MP@ML (SDTV) decoder. Table 31.4 shows the memory usage of the reduced resolution decoder in comparison to an SDTV decoder for reference. The following discussion is going to get pretty technical. For a review of the operation of MPEG-2 video coding and decoding, see Part 3 – Chapters 8 through 12 – for a five-part treatment of the subject.

The primary modifications to an MP@ML decoder are the addition of a bitstream pre-parser at the input, an up-sample function before and a down-sample function following the half-pixel interpolator in the motion compensation prediction path, extended registers in the syntax parser and variable length decoder, a simple row extension to the inverse discrete cosine transform (IDCT) function, and a down-sample function following the IDCT. The net result of these additions is the ability to decode an SDTV image from

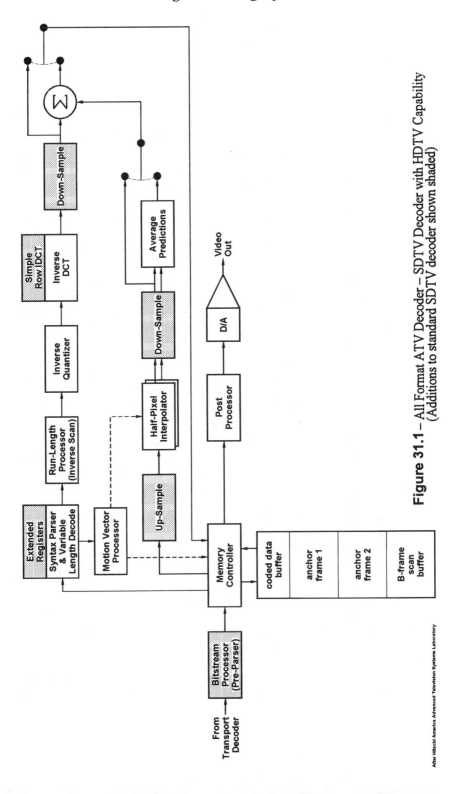

Figure 31.1 – All Format ATV Decoder – SDTV Decoder with HDTV Capability (Additions to standard SDTV decoder shown shaded)

After Hitachi America Advanced Television Systems Laboratory

an HDTV stream using the same memory as in a standard MP@ML decoder and with the same peak processing speeds as in an MP@ML implementation.

Looking at this in a little more detail, the pre-parser operates on the input bitstream to perform triage (a neat choice of terminology by the folks at Hitachi – the dictionary says it's "a system used to allocate a scarce commodity only to those capable of deriving the greatest benefit from it") on the incoming data, discarding less important coding elements, in particular high frequency DCT coefficients. It does this while the coefficients are still variable-length encoded and in the form of a run length of zeroes combined with the amplitude of the next non-zero coefficient. The result is that, by comparison to an MP@HL (HDTV) decoder, the pre-parser discards sufficient data to allow a smaller channel buffer to be used without overflowing while also reducing the channel buffer bandwidth requirements, and it discards a large proportion of run-length/amplitude symbols thereby allowing for simpler real-time syntax parser (SP) and variable-length decoder (VLD) functionality. These results are quite apparent in the upper half of Table 31.3.

In terms of the picture information that the pre-parser sheds, it is essentially the high frequency picture detail in both the horizontal and vertical directions that would be outside the bandwidth of an SDTV pixel matrix anyway. If we remember that the DCT function basically takes the picture in an 8x8 block of pixels and transforms it into eight horizontal and eight vertical sub-bands, it should be apparent that only the sub-bands from an HDTV image clustered around the DC coefficient should be of any use on an SDTV display. The pre-parser performs this triage by limiting the peak number of bits per macroblock by about a 6:1 ratio while reducing the 18 Mbits/sec average bit rate of the video stream to 12-14 Mbits/sec. The end result is a channel buffer bandwidth of the same 23 MBytes/sec as for MP@ML.

The pre-parser also controls the channel buffer fullness in the decoder to make sure that it will never overflow. This function, normally controlled at the encoder, is accomplished by making the pre-parser more aggressive in its dropping of coefficients as the buffer fullness rises.

Because so many of the DCT coefficients are eliminated, either in the pre-parser or in the inverse quantization (IQ) and IDCT processing, the computational requirements of the IQ and the complexity of the IDCT are both substantially reduced from that required for full HDTV processing. Just 10 or 11 of the 64 DCT coefficients are retained, and the MP@ML 8x8 IDCT function must be enhanced only slightly to accommodate the manner in which the the coefficients are eliminated. Since a processing engine is typically used for one dimensional IDCTs in MP@ML decoding, it would be just fast enough to do the column IDCTs on three columns of the HDTV image blocks. The row IDCTs must be handled separately, but only three columns are active in each row. This leads to the simple row IDCT function shown as the addition to the IDCT in the block diagram. Following the IDCT, a down-sampling (filtering and sub-sampling) function is performed to reduce the size (resolution) of the image stored in memory and output from the decoder.

One of the other requirements to perform the downconversion from HDTV to SDTV is to utilize the motion vectors of the HDTV image to perform the motion compensated prediction in a manner that will be useful with the spatially larger blocks (relative to the total image area) of the SDTV DCT coefficients. This is done by up-sampling (through interpolation) the stored I-frame and P-frame images (anchor frames) so that they fall on the same pixel matrices as were used in derivation of the motion vectors originally. Once the proper half-pixel interpolation of the the images is completed using the motion vectors, the displaced images are restored to the reduced resolution condition expected elsewhere in the MP@ML decoder by filtering and down-sampling.

RESULTS TO DATE

As mentioned earlier, the results of applying an all-format decoder to an HDTV bitstream were publicly demonstrated at the recent SMPTE conference. At those demonstrations, some knowledgeable viewers noted certain artifacts that they deemed unacceptable. (Note that I don't go by just what I saw because I have, in the past, demonstrated particular sensitivity to a wide range of artifacts and image distortions. I would be considered an "expert viewer.") They generally noted, however, that with a certain amount of improvement in those specific artifacts, the images would probably be quite satisfactory for non-expert viewers.

At a private demonstration at Hitachi in recent weeks, substantial improvement had already been achieved through the use of image post-processing. It should be noted that the images being employed for the development and demonstrations have been chosen because they are particularly susceptible to the processing being done. They contain large areas of complex detail and include significant motion. So there is good reason to believe that for most images, viewed on small to moderate size displays at normal television viewing distances, quite reasonable results eventually can be attained. It may always be the case that for large screens and closer relative viewing distances, HDTV decoders will be required. They are, after all, the reason that HDTV was developed in the first place.

FUTURE ENHANCEMENTS

As a final note, there may be a possibility that the image performance of all-format decoders can be improved (artifacts reduced) through the use of certain practices at the encoder. These practices would fit within the confines of the MPEG-2 standards and might include using shorter group of pictures (GOP) structures or using a higher proportion of B-frames to P-frames within a GOP, for example. Because of the down-sampling that goes on in an all-format decoder of the type described, distortions occur that propagate from one

P-frame to the next until the occurrence of the next I-frame. These distortions become progressively more visible until the process resets at the I-frame. This can be avoided by limiting the number of P-frames between I-frames to one.

Something to think about: Suppose the all-format decoders from one manufacturer benefit from one encoder practice while another manufacturer's devices require another, incompatible practice for best performance. If you want to help all-format decoders achieve better results, which one do you favor and how do you decide? Perhaps we will need another industry recommended practice to deal with this issue.

PART 10

Widescreen 525

The history of television technology has been marked by advances from one level to the next through compatible modifications and improvements. Thus it was never necessary to completely replace entire systems or facilities and start over (with one notable exception – the color wheel scheme – that failed). In moving to Advanced Television, a continuation of the use of compatibility to ease the process makes a great deal of sense, particularly economically.

One way to make a compatible transition in the studio is to use much of the existing equipment and infrastructure to produce widescreen (16x9) images and to post produce and distribute them. Such material would fit right into the SDTV domain described in the last part, or it could be used to upconvert to HDTV raster structures for inclusion in HDTV programming. Developing these techniques was a personal activity during the late '80's and very early '90's. So it was with some personal pleasure that the adoption of the proposals by a number of manufacturers at the 1993 NAB Convention was recognized This led to a two-article discussion of what I've called Widescreen 525 that appears in Chapters 32 and 33.

Note the absence of use of the term SDTV in these chapters. It did not exist at the time they were written, although substantial attention was given to describing what it is all about. Its use did not come into vogue for at least another year.

Since then, SMPTE has adopted standards for both possible sampling rates that can support Widescreen 525 in the studio, facilities have been built that can feed Widescreen 525 signals into digital video compression systems, and set top decoders are in the hands of consumers that can decode those signals and pass them on to appropriate displays. Only the displays themselves are missing from the marketplace in any volume (in the U.S. — they are very popular in Europe and Japan).

32

Widescreen 525
as an Interim Step

A very significant trend became quite apparent at the recent [April, 1993,] convention of the National Association of Broadcasters. That trend is acceptance of the concept of using a widescreen, 525-line image for many applications in the future. Those applications range from an interim, albeit long term, step to true HDTV production to a permanent method for feeding compressed digital video multicasting systems. What makes the trend so significant and apparent is its demonstration, both on and off the exhibit floor, by a number of the major players among equipment manufacturers, using equipment that is closer to production hardware than it is to prototype.

Concurrently, there is a trend in the consumer electronics industry and in the non-broadcast distribution media to provide widescreen, 525-line programming to consumers. This is underscored by the demonstration at the NAB Convention of widescreen receivers by both Thomson Consumer Electronics (which manufactures the RCA, GE, and ProScan lines) and by Philips Consumer Electronics Company (which owns the Magnavox, Sylvania, and Philco brands). Widescreen receivers are currently on sale in Los Angeles and will be available nation-wide by the end of this year. Similarly, the Compressed Digital Video (CDV) systems being developed for use by DirecTV for Direct Broadcast Satellite (DBS) and for use by various cable operators who are moving to multicasting will all be capable of widescreen operation.

First published June, 1993. Number 11 in the series.

The impact of these trends is likely to be quite far-reaching in everything from future studio operations to future distribution systems. So, this chapter, we will look at the implications and the technology of what I have labelled Widescreen 525.

WIDESCREEN 525 DEFINED

Widescreen 525 was first described by this writer in a paper given at the SMPTE Television Conference in February, 1989. This was followed up with papers at several additional SMPTE conferences and NAB Conventions and with demonstrations of the technique at the NAB Conventions of 1989 and 1990. The demonstrations showed how Widescreen 525 could be used as a low cost, moderate performance entry method for production for HDTV distribution applications. Subsequently, it has been recognized by various sectors of the industry that the use of Widescreen 525 by itself yields sufficient improvement over today's television to be a worthy step on its own.

Essentially, Widescreen 525 is an extension of the equipment and systems currently in use in order to provide wide aspect ratio images with somewhat improved resolution. This makes use of the bandwidth overhead that is currently built into most systems beyond what can be transmitted to viewers. Widescreen 525 can be implemented at equipment costs that are the same as or only a few percent higher than those of current, comparable gear, depending upon the implementation choices made, as we will discuss shortly.

It should be noted that we are consistently avoiding use of the term Widescreen NTSC in this discussion. This is because the applications that will utilize the widescreen signals are all component in nature, and they are likely to be significantly degraded in performance by use of a composite input. This results from the effective noise added to the image by the imperfect removal of the subcarrier in the decoding that would be necessary for use of widescreen NTSC. Widescreen NTSC can be considered an alternative for lower performance applications, but we will limit our discussion to component signals in the widescreen form.

The fundamental theory of Widescreen 525 is alteration of the horizontal scanning and related functions, leaving virtually everything else in the system intact. This makes the image wider, yielding a 16:9 aspect ratio, with the same number of lines in the raster. In practice, in cameras, both the horizontal and vertical dimensions must change. We will examine such practical considerations later. For now, we can consider that the height stays constant, and the width grows wider to move from 4:3 to 16:9. This is shown in Figure 32.1.

THE WIDESCREEN RASTER

We can expand the picture horizontally by simply scanning the pickup device in a wider dimension horizontally as compared to the vertical dimension. When we do this, the amount of time available for the horizontal scan does not change because the field rate, number of lines, and blanking periods have not changed. Since a 16:9 image is one-third wider than a 4:3 image, this means that the same 52.5 μsec. must

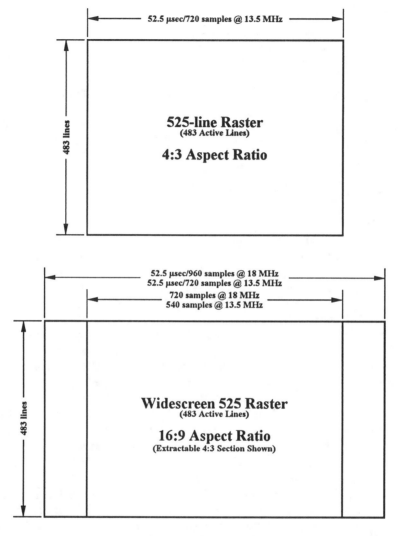

Figure 32.1 – Standard & Widescreen 525 Geometry

either carry one-third more information (i.e. bandwidth) if the resolution is to remain constant, or the resolution must be reduced to ¾ of its 4:3 value if the bandwidth is to remain constant.

Another way to look at the relationship between the aspect ratio, the resolution, and the bandwidth is to consider the general case of the horizontal resolution that can be achieved *versus* bandwidth for each of the two aspect ratios. This relationship can be seen in Table 32.1. The resolution is given in lines per active picture height (l/aph), which is a measure of the resolution that would be achieved horizontally in a space equal to the active height of the picture (discounting vertical blanking). This

Table 32.1 – Bandwidth *Vs.* Resolution

Bandwidth	4:3 (l/aph)	16:9 (l/aph)
4.2 MHz	331	248
5.0 MHz	394	295
5.75 MHz	453	340
6.0 MHz	473	354
7.67 MHz	604	453

measure allows expression of the resolution in a form independent of the aspect ratio, an important consideration when examining aspect ratio effects. (The values in Table 32.1 are based on precise values that approximate 80 l/aph/MHz for 4:3 and 60 l/aph/MHz for 16:9.)

DIGITAL IMPLEMENTATION

One of the important enablers for Widescreen 525 is the availability of digital equipment at about the same cost as analog. Making this practical is the use of the relatively new Serial Digital Interface (SDI) standardized by SMPTE a couple of years ago. The SDI allows systems to be built using coax cable interfaces between equipment, making digital system cost and complexity about the same as for current analog installations.

Digital equipment provides greater bandwidth overhead than typical, equivalent analog equipment, and this makes it easier to provide for the wider image. To take one example, NTSC transmission has a bandwidth of 4.2 MHz., which gives a horizontal resolution of about 330 lines, as can be seen in Table 32.1. Systems using digital component signals (based on the SMPTE 125M and SMPTE 259M standards), typically employing D-1, soon-to-be D-5, or Digital BetaCam recorders, have a bandwidth of 5.75 MHz. This is based on their use of 13.5 MHz sampling and 720 active samples per line, with an appropriate filter characteristic. Table 32.1 shows that this results in a resolution of about 340 lines in a 16:9 image format, roughly the same as current NTSC broadcasting.

It should be noted that the same digital component equipment and systems provide a resolution of a bit over 450 lines in their current 4:3 applications. It may be desirable to maintain this resolution in future applications. If so, it is necessary to increase the bandwidth by 4/3 (the ratio of the image widths) to keep the resolution constant. This results in a bandwidth of 7.6667 MHz that becomes available if the digital sampling rate is increased to 18 MHz, with 960 active samples per line and a proportional filter characteristic.

Since the Serial Digital Interface will certainly be the implementation method of choice in the future, it is important to relate these sampling schemes to the SDI data rate. SMPTE 259M provides for transportation of 10-bit samples across the interface. At 720 samples per line and 13.5 MHz sampling, this results in 270 Megabits per second (Mb/s) across the interface. At 960 samples per line and 18 MHz sampling, the data rate becomes 360 Mb/s. It is support for this latter value that has led most digital routing switcher and distribution equipment designs to

Table 32.2 - Summary of Component Digital System Characteristics

Samples/ Active Line	Sample Rate	Bandwidth	4:3 (l/aph)	16:9 (l/aph)	Data Rate (10 bits)
720	13.5 MHz	5.75 MHz	453	340	270 Mb/s
960	18.0 MHz	7.67 MHz	604	453	360 Mb/s

handle approximately 400 Mb/s. More on this later. The relationships of the sampling rates, bandwidths, active samples per line, and data rates are summarized in Table 32.2.

CHOOSING THE RATE

One of the hottest debates about Widescreen 525 for the last year or so has been which sampling/data rate to use. When originally proposed, the goal was to maintain the resolution that was available from 13.5 MHz sampling in 4:3 applications. This meant moving to 18 MHz sampling for Widescreen 525 implementation. An important factor in this was consideration that it will be necessary to extract a 4:3 image from 16:9 production in many applications, for example, for simulcast NTSC broadcasting, or for display on older televisions in viewers homes when 16:9 is delivered all the way to the home over CDV multicasting systems.

The resolution in an extracted 4:3 image, in lines per active picture height, is the same as the resolution in the wider image from which it is extracted. Thus if 720 samples/line and 13.5 MHz sampling were used for 16:9 images, an extracted 4:3 image would have only 540 samples per line and 340 l/aph. This is just enough to match current NTSC transmissions, but is considerably less than achieved in current component post production facilities. This led to the push for 18 MHz implementation.

There was thought to be a cost penalty for moving to 18 MHz of about 10-15 percent. This was deemed to be reasonable for a 33 percent increase in performance. But then it was recognized that, for facilities already operating at 13.5 MHz, the cost penalty would be much higher. This results from the fact that with relatively minor modifications they can operate their current equipment in 16:9 mode at 13.5 MHz. 18 MHz operation would require almost total replacement of the current equipment complement. For operations already utilizing component digital equipment, this makes 13.5 MHz operation far more attractive.

The question was also raised as to how visible the difference really is between 13.5 MHz and 18 MHz sampling. It has been appreciated for some time that the proportion of images with a given amount of detail grows smaller as the level of detail goes up. This means that less benefit is achieved for higher resolutions unless screen size is increased markedly. Some manufacturers have done private testing and have found that, at least on moderate size screens (e.g. 27-inch), the difference between 13.5 MHz and 18 MHz sampling is noticeable on only a small proportion of the images they have tried. What happens on larger screens, especially after line doubling, is still an open question. The sampling rate issue is certain to continue to dominate the discussion of Widescreen 525 for some time yet.

IMPLEMENTING WIDESCREEN 525

The theory of Widescreen 525 is relatively simple. So, too, is its implementation, although there are specific considerations that must be given to different types of equipment. We will take a look at cameras, recorders, distribution equipment, switchers and digital effects devices, and graphics equipment. That discussion appears in the next chapter.

33

The Practical Side of Widescreen 525

In the last chapter, we began looking into the use of 16:9 aspect ratio imaging combined with all the other characteristics of standard 525-line television – an approach we called Widescreen 525. (Note that all the comments made about Widescreen 525 apply similarly to Widescreen 625.) Conceptually, this simply involves the scanning, or reading out, of a wider picture in the same time normally taken for traversing the width of a 4:3 image. In practice, things are just a little more complicated than this.

This time, we will look into what is actually involved in implementing Widescreen 525. This includes the possibilities of modifying existing equipment as well as of buying new gear, and it includes the impacts of choices made about the sampling rate or bandwidth to be used. Our objective is to lay the groundwork for design and purchasing decisions to be made now and into the foreseeable future. As explained last chapter, we will limit our discussion to component operation because composite (NTSC or PAL) operation is too limiting in the performance obtainable from compression systems further downstream.

IMPLEMENTING WIDESCREEN 525

The theory of Widescreen 525 is relatively simple. So, too, is its implementation, although there are specific considerations that must be given to different types of equipment. We will take a quick look at cameras, recorders, distribution equipment, switchers and digital effects devices, and graphics equipment as representative of the

---

First published July, 1993. Number 12 in the series.

entire range of television equipment. The discussion to follow is not intended to be exhaustive but should give a flavor of the kinds of things that must be addressed.

There are two ways to approach Widescreen 525 implementation. One is to design and build new equipment specifically for the purpose. The other is to modify existing 4:3 equipment to permit widescreen operation. There is also the matter of implementation with 13.5 MHz sampling or with 18 MHz sampling, as we discussed last chapter. This discussion will not assume any of these approaches exclusively but will speak to each as appropriate.

SAMPLING RATE

The sampling rate issue has a significant impact on all the other factors in the implementation of Widescreen 525. If the resolution of current 13.5 MHz sampling is to be maintained in the widescreen mode, then the sampling rate must be stepped up to 18 MHz (1/3 higher, to account for the 1/3 wider image). This offers the possibility to deliver to viewers over 1/3 more resolution than NTSC does today when Compressed Digital Video (CDV) delivery systems are in place in cable and DBS operations in a few years. If broadcasters are permitted to transmit CDV programming and not just HDTV, they, too, could gain the 1/3 resolution benefit. In addition to providing the higher resolution now, the use of the wider bandwidth may also provide a transition path to HDTV operation later by permitting use of a lightly-compressed (in the range of 3:1 to 4:1) HDTV signal with the equipment installed now.

The choice of 18 MHz sampling will require that almost all the equipment in the installation be new, since there are only a few types of equipment currently in the marketplace capable of working at the higher bandwidth (largely routing and distribution equipment that operates up to 400 Mb/s, about which more later). Existing equipment all operates at a 13.5 MHz sampling rate. There is very little possibility to modify 13.5 MHz equipment to operate at 18 MHz.

In contrast, it is generally relatively easy to modify equipment with the current 13.5 MHz sampling to operate in a widescreen mode. This usually takes the form of changing software or firmware that determines the shapes of patterns such as wipes and other effects. Analog equipment also can be relatively easily modified to handle the wider images in a similar manner. Both of these types of changes were demonstrated at the 1989 and 1990 NAB Conventions in technology exhibits organized by this writer.

Selection of 13.5 MHz sampling will allow virtually all existing digital component production and post production equipment to continue to be used. This will provide a very significant cost savings *vs.* the cost of re-equipping with new equipment for those facilities (largely production houses) that have already installed significant facilities built around 4:2:2 digital operation (at 13.5 MHz sampling). It will mean, however, that they will be limited in the resolution they deliver to approximately what can be done with NTSC today. (See Chapter 32 for more on this.) Now, ... to look at specific categories of equipment:

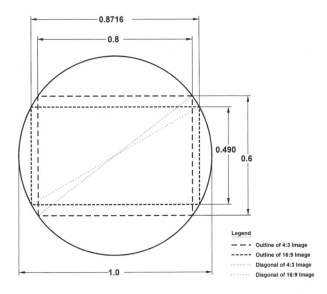

Figure 33.1 – Geometry of Camera Optics for 4:3 & 16:9 Aspect Ratios

CAMERAS

While the theory is that the Widescreen 525 image just gets wider relative to a 4:3 image, making that type of change in practice would require rather significant modifications to the optical portions of camera systems. Instead, it is preferable to maintain a constant size for the sensor and the related optics. This requires that the diameter of the optics and the diagonal of the sensor be matched and utilize a standard size. For our examples here, we will assume 1-inch optics. This will help in seeing the relationships more easily. To get to the more practical sizes of ¾-inch or ½-inch optics, just multiply all dimensions by ¾ or ½, respectively.

Figure 33.1 shows a circle representing the outline of the optical system. Inside that circle are two rectangles that have a common diagonal length. One represents a 4:3 image with a diagonal equal to the diameter of the circle. The other represents a 16:9 image with a diagonal equal to the diameter of the circle. As can be seen, in order to keep the diameter/diagonal constant and equal to the optics, it is necessary both to increase the width of the image and to reduce the height. Thus practice, at least in cameras, differs a little from the theory of just widening the image.

It is possible to convert existing cameras to widescreen operation. If they are tube cameras, this only requires adjustments along the lines of what is shown in Figure 33.1. (The width is increased to 108.95 per cent of its original 4:3 length, and the height is decreased to 81.71 per cent of its original 4:3 dimension. If possible, the bandwidth should also be increased.) If they are CCD cameras, there are two possibilities. One is to use an anamorphic lens that squeezes the image horizontally so that, at the output of the lens, the image has ¾ the width it has at the input of the lens. The image height would remain the same. These have been shown, at least in prototype form, on a number of occasions. The other possibility is to replace the CCD block with one having different sensors. The geometry of these new sensors would be the same as shown in Figure 33.1 for tube sensors using the same diameter optics. A number of manufacturers are planning to offer this possibility, at least for their newer cameras. Comments from several of the them put the price for doing this in the range around $15,000.

Of course for new camera designs, there are opportunities to do some additional things. For example, it is quite possible to make the camera switchable between 16:9 and 4:3. This can be done by always scanning 16:9 and throwing away the outer pixels on each line so that the center 4:3 image is left. The center portion must then be re-timed in a buffer to fill the interval between the end of one blanking period and the beginning of the next.

There are also possibilities to improve the resolution of the imager in new designs, so that the image is oversampled, and, when combined with a preceding optical filter, aliasing is reduced. This is already done in many current 4:3 cameras. One manufacturer that plans to use 13.5 MHz sampling as the basis of its overall system is contemplating using 18 MHz sampling at the sensor to provide such oversampling. This is an elegant solution if 13.5 MHz is enough in the remainder of the system. For overall systems based on 18 MHz sampling, oversampling horizontally by the same proportion would require on the order of 1280 pixels and a 24 MHz initial sampling rate.

RECORDERS

Recorders really have no knowledge of whether the aspect ratio they are recording and playing back is 4:3 or 16:9, or anything else for that matter. They do no processing that is dependent on image geometry or content. Thus any component recorder with adequate bandwidth can be used for widescreen applications.

For existing analog recorders, bandwidth will be the only issue. In the widescreen demonstrations conducted in 1989, modified M-II recorders were used with their luminance bandwidths increased to about 6 MHz. The color difference bandwidths were left unchanged at a little over 1 MHz. More would be desirable for high quality post production. Wideband 625-line Betacam recorders with 8 MHz luminance bandwidth intended for widescreen applications have also been delivered to a British DBS operation.

Digital recorders generally provide wider bandwidths than standard analog machines. Machines based on the 4:2:2 level and 13.5 MHz sampling of the SMPTE 125M and SMPTE 259M standards, such as D-1, D-5, and Digital Betacam machines, all provide a 5.75 MHz bandwidth. Similar bandwidth will result from the combination of D-2 machines and the new coefficient recording adapters that allow them to record and play compressed component signals. In addition, the D-5 machines have a switchable mode that allows them to record

18 MHz sampled 4:2:2 signals at 8 bits. (They normally record 10 bits at 13.5 MHz sampling.)

DISTRIBUTION AND ROUTING EQUIPMENT

Distribution and routing equipment is among the easiest to make usable in any of the modes we have been discussing. Again, the important characteristic is bandwidth. For analog distribution and switching, most distribution amplifiers and routing switchers provide a minimum of 10 MHz bandwidth per channel, with many good to 30 MHz or beyond. Of course for analog systems, the distribution and routing of component signals are handled with three channels in parallel.

Digital distribution and routing can use either parallel or serial interconnections. The parallel interface requires a cumbersome DB-25 connector with at least 9 and preferably 11 carefully balanced pairs. The cumbersome nature of this interface has been one of the factors holding up widespread digital implementation.

Serial digital interfaces, on the other hand, utilize the same coaxial cables as used for analog video, with the same BNC connectors and the same patch panels. Combined with their ability to handle signals ranging from the 143 Mb/s of digital NTSC, through the 177 Mb/s of digital PAL, the 270 Mb/s of 13.5 MHz sampled components, to the 360 Mb/s of possible 18 MHz sampled widescreen components, serial digital interfaces are becoming a major enabler that is helping digital technology burgeon.

Since the idea of operation at 360 Mb/s came on the scene, most manufacturers of distribution and routing equipment have made operation of their equipment to 400 Mb/s more or less the standard. This way, whatever happens, the equipment will serve its intended purpose. There are some issues that need attention, however, in pushing systems to the upper ends of their operating ranges. These include the facts that usable cable lengths go down, that equalizers become more difficult and expensive, and that jitter becomes more of an impediment.

SWITCHERS AND DIGITAL EFFECTS

Production switchers and digital effects units can be divided into two sets of functionality they provide: the functions of routing and switching we have just discussed, and the functions of image processing. In the routing and switching sections, there is no dependence on image geometry or content. Bandwidth is the prime consideration, and the comments just made about routing and distribution apply.

In the image processing portions of switchers and digital effects devices, many functions are closely related to the geometry of the image. In these cases, changes must be made to provide the intended results when a wide aspect ratio is used. An example of this is in wipe pattern generation.

For instance, if a circle wipe is to look like a circle, the number of pixels that it is wide must be greater by one-third if the aspect ratio is 4:3 than it must be if the aspect ratio is 16:9. The exact number of pixels required depends on the size of the circle in terms of its height in the picture. Thus if the circle is 100 lines high, it must be about 112 pixels wide in a 4:3 image, and it must be about 84 pixels wide in a 16:9 image. This result follows from the fact that the 16:9 image, being wider yet having the same number of pixels per line, must cover more of a line with each pixel. So fewer are needed to cover the same space as in a 4:3 image.

Since the patterns being generated in switchers and digital effects units are often known shapes, the formulas or patterns used to generate them must be changed to account for the different aspect ratios. Sometimes this can be done from the control panel of the unit or can be programmed from an external control input. This was demonstrated in the exhibitions mentioned previously, where a circle wipe was used and the switcher was changed on a vertical interval switch from 4:3 to 16:9 operation or back.

In digital effects units, similar concerns arise where patterns are generated. The more complicated manipulations of rotating pictures or generating page turns may require knowledge of the aspect ratio. Often these manipulations are not adjustable through use of an aspect ratio control, and the differences must be hard coded into firmware. Attention must be paid not only to the manipulation itself, but the key signals fed out to switchers must be considered as well.

GRAPHICS EQUIPMENT

Graphics equipment includes character generators, paint systems, and the like. Whether there are any differences required in these units beyond bandwidth considerations depends upon whether known shapes are being generated. For example, in generating the letters of the many fonts used in character generators, there are certain expectations about what various characters of the alphabet should look like. Presenting them one-third wider in a 16:9 aspect ratio image would not be acceptable. Thus they would have to be altered to take one-third fewer pixels horizontally when presented in 16:9.

When images are generated in a free form manner, as in a painting activity, knowledge of the aspect ratio is not required. In fact, the aspect ratio is factored into the image by means of displaying the image in the correct aspect ratio on the monitor used by the artist in creating the image. Any known shapes used in creation of the image may require adjustment, however. Our circle is a good example. If the function is the generation of a circle by the setting of the center point followed by setting a point on the circle itself, the artist then expects to see a circle, not an elipse.

By now, the principal issues in using a 16:9 aspect ratio in a 525-line environment should be quite apparent: bandwidth throughout the system and its equipment, and known shapes where patterns are generated in image processing, manipulation, or creation. Widescreen 525 and 625 seem like techniques that are quite likely to have a place in our futures. Our discussion should show that their implementation is a natural extension of what we have already been doing. A little planning and thought will go a long way to making the transition to widescreen operation a smooth and easy one.

PART 11

Other Matters

Advanced Television is such a broad canvas on which to paint that there are always lots of absorbing and useful topics to illustrate. Sometimes they do not fit into the mainstream of the subject matter we cover, but they deserve attention nonetheless. Two such topics are data broadcasting and jitter in serial digital signals. Since they do not fit particularly well into any of the other major subjects in this book, they are grouped here under the rubric of Other Matters.

Data broadcasting, in the sense used here, is the inclusion of high speed data subchannels within the spectrum of ordinary NTSC television signals. The technology to support such transmissions is the same technology as is used for purely digital transmission in the Grand Alliance system and in the many cable and similar digital video compression delivery systems under development. Thus it is most appropriate to find these techniques included in a book on

Advanced Television. The only difference in the use of the technology is that the signals are structured to fit within a host NTSC signal, ideally without interfering with the content of the host.

Two data broadcasting systems have been under consideration by an industry committee looking to standardize one or both. The WavePhore system is considered in Chapter 34. The Digideck system is covered in Chapter 35. In addition, Chapter 35 describes an application for data broadcasting that currently makes do with the old teletext method for transmitting data signals within a host NTSC signal and that could make good use of the wider data pathways that might be provided by either or both of WavePhore and Digideck.

Jitter is something that afflicts all serial digital signals. The extent to which it does so determines how far and under what conditions a serial digital signal can be delivered from a transmitter to a receiver. This is true whether the signal propagates through the ether or over a cable. It is true whether the signal is a relatively low speed digital audio signal or a very high speed studio HDTV serial digital signal. Jitter will affect all of the packetized signals and all of the modulated streams discussed throughout this book.

With such a universal effect upon signals of the sort used in Advanced Television applications, it is fitting that we conclude our time together here with consideration of jitter in Chapter 36. We look at how it is defined, how it is measured, how it accumulates, how it can disrupt, and how it is cured.

34

Data Broadcasting — WavePhore

The idea of broadcasting data along with normal NTSC transmissions has been around for a long time. Back in the late 1970's and early 1980's, a system called teletext was developed. It did not catch on in North America, but it has been in service in Europe ever since. There were three main reasons it didn't work here: (1) it didn't produce acceptable data recovery in enough locations, (2) there wasn't a single standard that could be fully developed by the industry, and (3) the program offering was not one that either the public or advertizers were willing to support financially.

A lot has happened in the intervening years. The world is going digital. Personal computers have become widespread, both in businesses and more recently in homes. There are many new applications looking for pathways to deliver their data to the public or subsets of the public. Terms like Internet, Information Superhighway, and National Information Infrastructure have joined the common lexicon. And there are now technical solutions that make it far more likely that digital signals can be accurately recovered in a wider range of locations than used to be possible.

Efforts are currently under way to select a technical approach for high rate data broadcasting from a field that has already been narrowed to two candidates. The work is being done under the auspices of the National Data Broadcasting Committee (NDBC), formed jointly by the Electronic Industries Association (EIA) and the National Association of Broadcasters (NAB). In

---

First published July, 1995. Number 30 in the series.

addition, there are a number of schemes that transmit data in the vertical interval of NTSC signals at somewhat lower data rates. We'll look at three data broadcasting systems in two parts, this chapter and next — the two that are contending for NDBC honors and an example of the vertical interval variety of system.

NDBC EFFORT

The NDBC was established in 1993 as an industry vehicle to select a single method for transmitting high rate data as an ancillary service to television broadcasting. It issued a Request For Proposals (RFP) that brought a number of responses from system proponents. The review process within the NDBC resulted in reduction of the number of systems under consideration to two – those of WavePhore and of Digideck.

Programs of both laboratory and field testing were defined by the NDBC, and laboratory tests were conducted at the Advanced Television Test Center (ATTC) in December, 1994. At a meeting on April 28, [1995,] the NDBC approved the Digideck system for field testing on the basis of the laboratory test results but declined to so approve the WavePhore system. The NDBC plan was to begin field testing of the Digideck system on about June 15, but that date seems highly optimistic and unlikely at the moment.

Meanwhile, WavePhore has indicated that it learned a great deal from the laboratory testing (about which more shortly) and contends that it can correct the deficiencies in its system found in that lab testing. At a meeting of the NDBC on June 2, it was agreed that WavePhore would be given another opportunity in the lab to see whether it merits going to field testing.

FCC NPRM

While all of this has been going on, the FCC adopted a Notice of Proposed Rulemaking (NPRM) on April 10, 1995, looking toward the establishment of rules to allow "digital data transmission within the video portion of television broadcast station transmissions." The NPRM defines two general categories of transmission methods – (1) the "overscan" approach in which data is carried on the active picture lines along with the image and is hidden from view at the sides, top, and bottom of the television raster in areas normally shielded by the picture tube mask and (2) the "sub-video" technique in which information is inserted in a manner that could affect the regularly viewable portions of the television picture but that nevertheless remain undiscernable by the ordinary viewer.

The FCC has had a number of proposals on methods for carrying ancillary data using each of the two categories. An example of the overscan method is the AMOL technique developed by A.C. Nielsen that uses line 22 to send data that identifies programs for audience measurement purposes; it has been in use under temporary authority from the Commission since 1990. There have been other proposals for various line 22 uses and for other methods of putting data in short pulses at the beginnings and ends of lines. These all use fairly straightforward technology and so need not concern this discussion.

Approval of the WavePhore system was sought from the FCC by WavePhore in a request for declaratory ruling in December, 1993. Its request has been folded into the current NPRM. The NPRM also specifically seeks information on the Digideck system as well as others that may impact the policy the Commission ultimately chooses to promulgate.

The NDBC decided at its April 28 meeting that it will submit comments on the NPRM. The approach it will take remains to be seen. It would be smart for the Commission to wait until the NDBC completes its work before proceeding with a decision. It won't get any better information than it will ultimately receive from the committee.

WAVEPHORE TECHNOLOGY

Turning now to the technologies of the two sub-video systems, the WavePhore system works essentially by cleaning out a portion of the NTSC video baseband and inserting a low level data carrier into that region. The system is called "TVT1/4" and uses the area from 3.9 to 4.2 MHz, as shown in Figure 34.1b. (A normal NTSC spectrum is shown in Figure 34.1a for comparison purposes.) The data carrier is injected at a low level, about 20 dB above the noise floor of the video signal on which it piggybacks.

The TVT1/4 nomenclature derives from WavePhore's intention to fit a T1-rate (1.544 Mb/s) data channel into the television signal. As the practicalities of the matter became evident, however, it was necessary to drop back to one-quarter of that rate (384 kb/s); hence the TVT1/4 appelation. In order to meet the need for more error correction pointed out by the lab testing, it may be necessary to drop the data rate further, to about 305 kb/s – about which more in a moment.

In TVT1/4, the lower sideband of a bi-phase modulated signal having a carrier frequency of approximately 4.197 MHz is filtered with a bandpass filter to preserve about 300 kHz of modulated data signal which is summed with the video signal. The carrier frequency is precisely an odd multiple of one-quarter of the horizontal line frequency so that the energy packets of the modulated data spectrum interleave with the spectra of the luma and chroma signals. This is similar to the manner in which the luma and chroma themselves interleave with one another through use of a half-line offset.

The data is structured so as to be synchronous with the carrier and, consequently, with the horizontal line frequency. Data is only transmitted during the active period of active lines in the picture. Thirty bits of data are transmitted per video line with a single, fixed-value bit transmitted at the beginning of the sequence on each line for carrier synchronization purposes. This is shown in Figure 34.2.

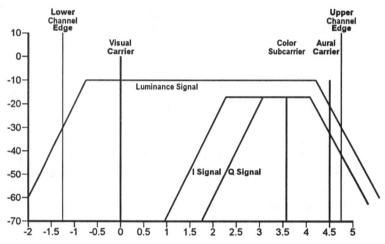

a) Normal NTSC RF Spectrum as Defined by FCC Rules
Roll off regions extended from minimum required slopes

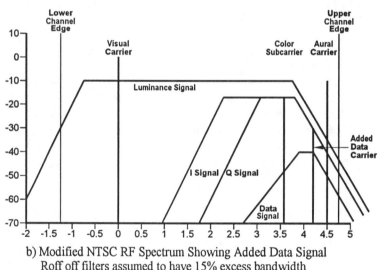

b) Modified NTSC RF Spectrum Showing Added Data Signal
Roff off filters assumed to have 15% excess bandwidth

Figure 34.1 – Normal NTSC Spectrum & Modification for WavePhore TVT1/4

With 240 lines used for data in each field, at 59.94 fields/sec, the raw data rate is 431.568 kb/s. The difference between this value and the throughput data rates mentioned earlier is the bit rate spent on error correction coding (ECC). The initial ECC applied by the system is the Reed-Solomon algorithm. The data is divided into segments of 205 bytes, and 20 bytes of ECC are added to each segment, creating a block with a total length of 225 bytes.

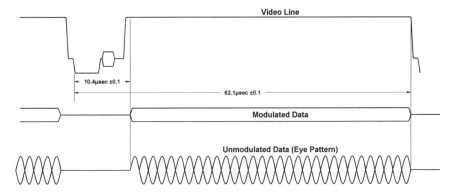

Figure 34.2 – Relationship Between Video Signal and Data

225 bytes (x8) represent 1800 bits. 1800 bits, when divided by the 30 bits per line, yield 60 lines per block, or 4 blocks per video field of 240 active lines. Thus each field can carry 7200 bits or 900 bytes of data including ECC.

WAVEPHORE SYSTEM

As shown in the encoder block diagram of Figure 34.3, the input data is first scrambled before the addition of the ECC. This is similar to the method used in the scrambling of data in the serial digital interface (SDI) of SMPTE 259M for composite and component digital video. The difference is that in the SDI it is done to eliminate DC and low frequencies to achieve better receiver clock recovery while in the TVT1/4 system it is done to minimize the visibility of any data patterns that might otherwise appear in the video.

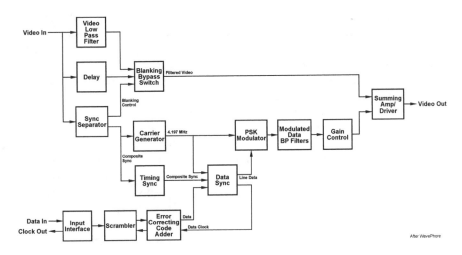

Figure 34.3 – WavePhore TVT1/4 Encoder Simplified Block Diagram

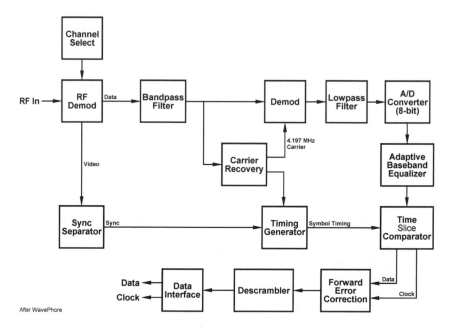

Figure 34.4 – WavePhore TVT1/4 Decoder Simplified Block Diagram

After scrambling, the data is synchronized with the video so that the correct number of bits are placed during each line interval. The data is then bi-phase modulated in the phase shift keying modulator using a carrier frequency derived from the incoming video signal. The modulated data is then bandpass filtered to extract the lower sideband, its gain is adjusted, and it is added to the video in the cleaned out region.

The input video, meanwhile, has been low pass filtered during the active line periods. During the blanking intervals, a blanking bypass switches to the output of a delay line that matches the delay time of the low pass filter so that the blanking period is not filtered but arrives at the correct time.

The WavePhore receiver essentially reverses all of the processes applied in the transmitter and delivers data and a clock on its output. Its block diagram is shown in Figure 34.4. Of course, it cannot restore the video signal information that was removed from the top end of the bandpass by the low pass filter in the encoder.

The major difference between the encoder processing and the decoder reverse processing is the addition of what WavePhore calls an adaptive baseband equalizer (ABBE). This is similar to the adaptive equalizers showing up in other types of advanced television systems such as the video echo cancellers that work with the Ghost Canceller Reference (GCR) signal for NTSC, the adaptive equalizer in the Grand Alliance ATV system and its progenitors, and the adaptive equalizers that will be in all of the cable and wireless cable digital set top boxes. The adaptive equalizer is probably the most important single development over the last decade or so that will make the difference between the early broadcast teletext attempts and the current efforts at data broadcasting.

WAVEPHORE TRADE-OFFS

It should be apparent that the filtering of the video to create space to insert the WavePhore digital carrier reduces the bandwidth and consequently the resolution of both the luminance and the chrominance signals of the underlying NTSC signal. The luminance bandwidth is reduced from 4.2 to 3.9 MHz. The bandwidth of the upper sideband of the chrominance is reduced from about 600 kHz to about 300 kHz. It is important to note that the limitation of the upper sideband to 300 kHz will mean that receivers cannot recover the I and Q signals at frequencies higher than this value, even though the lower sideband is still present, since both sidebands are required for quadrature demodulation. The results will be loss of color saturation and chroma crosstalk.

The trade-off for this reduction in chrominance bandwidth is that the data signal can be carried along with the video signal as it traverses the path from studio, to network, to station, to VTR, to home, and so on. It also requires no significant modifications to the exciter or transmitter for implementation. Instead, an encoder/inserter is used somewhere along the video path.

WAVEPHORE TESTING

Test results of the TVT1/4 system were certainly disappointing for WavePhore. They showed significant losses in video performance of the host NTSC signal in areas such as subjective video quality tests, chroma gain, group delay, and VHS recordability, where chroma shift, bleeding, and smear were noted. The TVT1/4 system also exhibited significantly lower data performance than the Digideck system in terms of the effects of random noise, impulse noise, immunity to co-channel interference, immunity to adjacent channel interference, and immunity to multipath.

WAVEPHORE IMPROVEMENTS

As a consequence of these test results, WavePhore has recognized some of the deficiencies of its own original testing and will make improvements to its system. The problems in its developmental testing come from having done all of its work in the field in a relatively isolated (from an RF point of view) place like Phoenix. There is essentially no co-channel interference in Phoenix, and the nearest adjacent channel is 90 miles away. Multipath is relatively low, as are the impulse noise sources that plague other markets. WavePhore was achieving transmission and reliably recovering data over distances of 70 miles in Phoenix. The lesson here seems to be that for development of new systems, both lab and field testing are needed, not just one or the other alone.

The improvements planned by WavePhore follow from identification of the use of video sync for deriving the timing of the data symbols as the principal contributor to the errors that were seen in data recovery during the ATTC testing. Many of the impairments applied to the signal clobbered NTSC sync. The improved WavePhore system will use a pseudo-random sequence in the data for its data synchronization. This is similar to the technique used by the Grand Alliance for synchronizing digital transmission of ATV.

In addition, WavePhore will add another level of Reed-Solomon error correction coding and extensive data interleaving to further improve signal robustness. This will exact a price in data rate – the reduction from 384 to 305 kb/s mentioned earlier. The data interleaving will spread the data over 60 fields, thereby minimizing the impact of any single or multi-block loss. The cost of this will be a longer time delay and acquisition latency through the system and a larger data buffer, since one second's worth of data will have to be buffered.

It is also planned to use a SAW filter for the low pass function that clears space in the video baseband. This will replace the L-C filter that was the likely source of many of the video performance issues. In particular, the SAW filter is expected to minimize the chroma smearing that was perceptible in a number of situations. The efficacy of these changes will then be demonstrated in the further lab testing to which the NDBC has agreed.

DIGIDECK

The Digideck approach to high rate data broadcasting is the addition of a subcarrier in the RF spectrum of a television channel. The system, called "D-Channel," requires clearing out some of the channel at RF in order to make room for the injection of the subcarrier carrying the data. The subcarrier operates at a relatively low level in order to reduce the interference it causes to either its host or to neighboring stations. We will look at the Digideck D-Channel system in more detail in the next chapter.

ONTV

OnTV is a new system about to go on the air in Pittsburgh for both technical and business development purposes. It is one of the first attempts at providing Internet World Wide Web-style browsing over television broadcast channels. Currently based on vertical interval transmission, it is representative of the new types of applications that could benefit from the data capacity that might be provided by the WavePhore and/or Digideck systems. We will look at OnTV in more detail in the next chapter, too.

35

Data Broadcasting —
Digideck

In our last time together, we began an exploration of the world of data broadcasting as an ancillary service to standard NTSC television transmission. We recognized the work of the National Data Broadcasting Committee (NDBC) in this area, saw that the FCC recently issued a Notice of Proposed Rulemaking (NPRM) on the subject, and examined one of the the two systems currently under NDBC evaluation – the WavePhore system.

This time around, we will look at the details of the other system being considered by the NDBC – the Digideck system – and we will look at a modern application for data broadcasting that is starting out using more conventional data transmission techniques. In the long run, it may well be a combination of the new transmission methods with some of the new applications that are now possible that ultimately will drive implementation of digital broadcasting in North America.

DIGIDECK TECHNOLOGY

The Digideck approach to high rate data broadcasting is the addition of a subcarrier in the RF spectrum of a television channel, as opposed to the baseband approach used by WavePhore that we investigated last chapter. The system, called "D-Channel," requires clearing out some of the channel at RF in order to make room for the injection of the subcarrier carrying the data. The subcarrier operates at a relatively low level in order to reduce the interference it

---

First published August, 1995. Number 31 in the series.

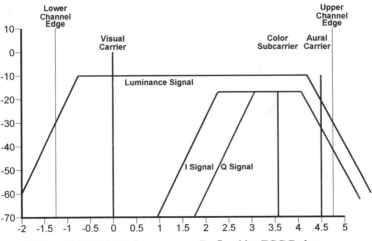

a) Normal NTSC RF Spectrum as Defined by FCC Rules
Roll off regions extended from minimum required slopes

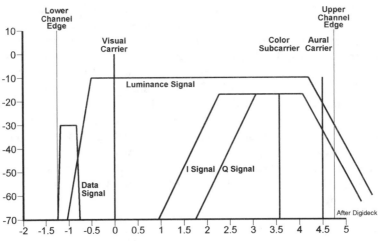

b) Modified NTSC RF Spectrum Showing Added Data Signal
Roll off filters assumed to have 20% excess bandwidth

Figure 35.1 – Normal NTSC Spectrum & Modification for Digideck D-Channel

causes either to its host or to neighboring stations. See Figure 35.1b for the Digideck spectrum, and compare it to the WavePhore spectrum shown in Figure 34.1b in the last chapter. (A normal NTSC spectrum is shown in Figure 35.1a for reference.)

In D-Channel, the signal introduced to carry the data is modulated with differential quadrature phase shift keying (DQPSK). This creates a constellation with four points, as shown in Figure 35.2. The four-point (2^2)

modulation results in a bandwidth efficiency of 2 bits per Hertz of bandwidth. It is placed 1 MHz below the picture carrier, has a channel data rate of 700 kb/s, and occupies a bandwidth (-3 dB) of 350 kHz. With the 20 percent excess bandwidth required to account for filter skirts, it occupies ±210 kHz from its carrier frequency (420 kHz total) at the -30 dB points.

The new carrier is located one-quarter megahertz above the lower channel edge and is injected at a level 30 to 36 dB below that of the picture carrier at peak of sync (-30 to -36 dBc). Thus it is at least 60 dB down from the picture carrier 40 kHz above the lower channel edge, easily meeting FCC out of channel radiation requirements if there are no transmitter linearity issues (about which more shortly).

To make room for the new signal, the vestigial (lower) sideband (VSB) response of the visual carrier must be adjusted to provide a steeper slope. Normally, the slope starts at -750 kHz or so below the visual carrier and falls below -20 dB (relative to the low frequency response) by -1.25 MHz, the bottom end of the channel. To accommodate the D-Channel signal, the slope must be increased and its start moved higher in frequency so that it begins at -500 kHz and is down more than 40 dB by -800 kHz, as shown in Figure 35.1.

Reed-Solomon error correction coding (ECC) is applied to the data with 25 per cent of the channel data rate devoted to ECC overhead. Thus the 700 kb/s channel data rate yields a net data rate of 525 kb/s. The actual system throughput will depend upon a number of factors to be specified by the NDBC such as the length of packets, number of sync bytes, and such. In the system tested at the Advanced Television Test Center (ATTC), 51 data bytes were attached to 16 ECC bytes and one sync byte for a total block length of 68 bytes. With this arrangement, the forward error correction system in the receiver could detect up to 16 byte errors per block and correct up to 8 byte errors per block.

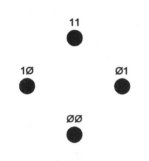

a) Ordinary QPSK Modulation
with bit values shown

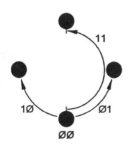

b) Differential QPSK Modulation
with bit values shown

Figure 35.2 – Ordinary &
Differential QPSK Modulation

DQPSK MODULATION

There are two basic kinds of QPSK modulation. One associates each of the four points in the constellation with the values of two bits (00, 01, 10, and 11).

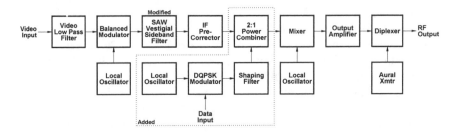

Figure 35.3 – Transmitter Modifications & Additions
for IF Insertion of D-Channel Signal

This requires some method of identifying at the receiver the correspondence between the specific phases of the signal and the bit values of each of the points. This is normally accomplished by periodically sending a known pattern to establish symbol synchronization. Doing so takes some data overhead and also increases the cost of the receiver for the necessary processing circuitry.

The other method of QPSK modulation works on a differential basis and so is called DQPSK. In DQPSK, instead of assigning specific values to the constellation points, the values are carried by the **changes** in the phase of the signal. Thus the same phase repeated in the next symbol has a value of 00. A change of +90 degrees represents 01. A change of -90 degrees represents 10. A change of 180 degrees represents 11.

This particular coding has several advantages. Errors in recovery of the data are most likely to occur when a constellation point is misinterpreted as one of the adjacent points (+ or - 90 degrees), rather than as the point all the way across the constellation. This is so because it takes less noise on the signal to push a constellation point into the detection box of one of its neighbors (a 90 degree phase error) than to completely invert the phase of the signal. The result is that only one bit will be in error in the data recovered from most errored symbols. (This is called Gray coding.)

Another advantage of DQPSK is that the complexity and cost of the receiver are reduced relative to those associated with ordinary QPSK. This results from the elimination of the need for symbol synchronization already discussed.

SUBCARRIER INSERTION

The altered vestigial sideband slope and location can be achieved in one of two ways, depending upon the transmitter system design. If the vestigial sideband filtering is done at IF, as is common with transmitters built in the last 20 years or so, the SAW filter (or other IF band-shaping filter) can be changed or an additional SAW filter can be inserted in the signal path to cut off the low end of the modulated spectrum. If the vestigial sideband filtering is done using RF plumbing on the transmitter output, the filter can be retuned to move the cut-off

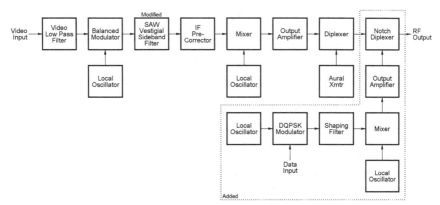

a) Addition of D-Channel Signal to Output of IF-Modulated Transmitter

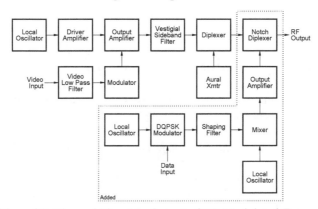

b) Addition of D-Channel Signal to Output of RF-Modulated Transmitter

Figure 35.4 – Transmitter Modifications for RF Addition of D-Channel Signal

point up in frequency. This has implications for how the new signal must be added to the transmitter output.

In the case of IF modulation, there are two possibilities for adding the data signal to the transmission. One is to add the signal at IF, after the VSB filter, and to amplify the signal along with the video through the transmitter's existing linear amplifier stages. This is shown in Figure 35.3. It requires that the transmitter have relatively linear amplifiers, although the fact that the subcarrier is down 30 or more dB should reduce the impact of amplifier nonlinearity. The other method is to add the subcarrier at the transmitter output as would be done with an RF VSB filter.

The cases of both an RF VSB filter and an IF VSB filter with output subcarrier addition are shown in Figure 35.4. For both setups, a low power amplifier is needed to supply the subcarrier signal for combining with the other transmitter output(s). Since the subcarrier is 30 dB down from the visual carrier, neglecting system losses, only one watt of output is needed for each kilowatt of visual transmitter output. To put this in perspective, a 15 kilowatt low band VHF transmitter would require a 15 watt subcarrier amplifier (plus enough additional to make up for combining losses), and a 120 kW UHF transmitter would need a 120 watt subcarrier amplifier (plus loss make-up).

Although the IF modulation case with IF subcarrier addition, the one in Figure 35.3, is the preferred method, some installations may require use of output subcarrier addition, as in Figure 35.4. This can potentially be accomplished by insertion of an additional notch diplexer in the transmitter RF output plumbing in a manner similar to that used for adding the aural transmitter output. For RF-modulated transmitters that use a vestigial sideband filter on the output of the video transmitter, the extra filtering could help in achieving the additional lower sideband suppression required.

ADAPTIVE EQUALIZATION

The receiver uses an adaptive equalizer to correct multipath, echoes, and the like. In the equipment tested, an equalizer with 32 taps was used. There are two taps per symbol period, making the equalizer 16 symbol periods long. This equates to approximately 46 microseconds at the 350 kilosymbol per second rate of the D-Channel system. Of this, between 40 and 45 microseconds are useful for echo cancellation after allowing for a few taps that must precede and follow the first and last occurrences of the signal, respectively. (Nyquist's limit must be obeyed in adaptive equalizers just as in A/D converters and elsewhere.)

Along with the normal form of adaptation to the channel characteristics, the adaptive equalizer is able to vary the amount of the filter that is used for pre-echoes and for post-echoes, rather than just using a fixed assignment of taps. This additional adaptation helps to optimize performance of the receiver with little cost impact.

DIGIDECK TRADE-OFFS

There are some trade-offs that result from the use of DQPSK as opposed to QPSK modulation. (Nothing comes without its price.) One is that errors tend to propagate from one symbol to the next; thus if one symbol is recovered in error, the next is likely to be as well. This means that the bit error rate (BER) at any particular operating level is likely to be double that for QPSK at the same level. The Gray coding prevents it from being any more than double. Since the impact of bit errors occurs on a logarithmic basis, the net result is a loss of only a decibel or so in the carrier-to-noise (C/N) threshold, depending on the particular operating point on the signal vs. error rate curve.

A different type of trade-off that occurs with the D-Channel system is that a separate path must be provided to get the data to the transmitter, as opposed to sending it along with the video. This separate path will be required from the point at which the data originates all the way to the exciter IF or the transmitter

RF output where the subcarrier is inserted. Techniques may be possible to carry the data along with the video through creation of some sort of side channel in the distribution system, but none has been proposed. If the data is unrelated to the video, of course, this is not so much of a problem, since the two are likely to require different paths, at least to the studio-to-transmitter link (STL) input anyway.

DIGIDECK TESTING

In general, testing at the ATTC went very well for Digideck. The D-Channel system achieved substantially unimpaired or only very slightly impaired results when evaluated subjectively for visible impairments to the host NTSC signal. Similar results were obtained when the audio performance of the host NTSC signal was subjectively evaluated.

With most of the standard video measurements, such as gain and group delay, luminance non-linearity, K factor, multiburst, and the like, only small, if any, reductions in performance were found. The biggest reduction in performance noted was a decrease in unweighted video signal-to-noise from a reference value of 54.5 dB to 46.2 dB. Since this would be beyond the threshold of visibility once weighting is applied, it should not be too significant.

With regard to the various interference cases, co-channel interference was no worse than that of ordinary NTSC, as was true for upper adjacent channel interference. Lower adjacent channel interference was only slightly degraded.

Bit error rate (BER) performance was measured at a BER of 1×10^{-5}. A carrier-to-noise ratio of 28 dB achieved this level of data performance. There was a cliff effect in data recovery, as one would expect. D-Channel showed good immunity to impulse noise, to co-channel interference, to adjacent channel interference, and to multipath.

The performance of the Digideck system in laboratory testing was such that the NDBC committee that evaluated the results recommended that D-Channel be approved for field testing, a recommendation in which the NDBC concurred. Field testing is still unscheduled as of this writing.

DIGIDECK IMPROVEMENTS

Despite its good performance in testing, improvements are still planned for the Digideck system. The most important change expected is the enhancement of the error correction through the addition of trellis coding at the transmitter and complementary decoding at the receiver. This could improve the threshold signal level for data recovery by 2-3 dB, although it comes at the cost of a reduced data rate. It may be possible to maintain the current data rate through a reduction in the amount of Reed-Solomon block error correction coding applied, but this will require a careful balance between the types of error correction practiced in order to achieve optimum performance.

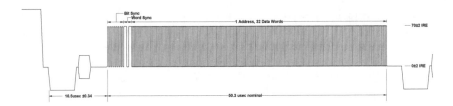

Figure 35.5 – Vertical Interval Data Transmission
with NABTS Amplitudes & Timing

ONTV

While the new techniques for data transmission are being developed, much can be done with more conventional methods when some of the newer supporting technologies are applied to them. In particular, new programming possibilities can be brought to the public that build on such recently popular applications as the Internet's World Wide Web. This situation is exemplified by the OnTV system that is about to launch in the Pittsburgh marketplace.

A partnership between public broadcaster WQED and the Television Computer Corporation, OnTV will initially deliver Web-style pages to data receivers in homes and businesses using the basic technology developed for teletext in the early '80's. These receivers pass the data on to personal computers of various sorts for further processing and for display.

What differentiates OnTV from the old teletext system is the use of Internet techniques to deliver a variety of data on a broadcast basis. It is also now possible to apply adaptive equalization (ghost cancellation) to the signal, if the application will support the cost, thereby overcoming one of the biggest problems experienced with teletext in the early '80's. (Your intrepid writer was responsible for one of the teletext tests back then, in San Francisco, where the success rate on installations was only about 25 percent because of the multipath and echoes in the Bay Area.)

For now, OnTV uses the North American Broadcast Teletext Specification (NABTS) data transport method to send data in the vertical interval. This approach puts a non-return-to-zero (NRZ) data stream on video lines as shown in Figure 35.5. The bit rate is 8/5 of the color subcarrier frequency, or 5.7272727272... MHz. With the NRZ coding, the highest fundamental frequency is half that, or 2.8636363636... MHz. The data is passed through a combination of a Nyquist filter and a low pass filter with a cut-off frequency of 4.2 MHz. The resulting spectrum is shown in Figure 35.6.

With the NABTS bit rate, it is possible to deliver 32 Bytes per line after subtracting the bits necessary for synchronization and addressing words. If one line per field of vertical interval is used, almost 7 MBytes per hour of raw data can be sent. Assuming that somewhere around half the data rate will be used

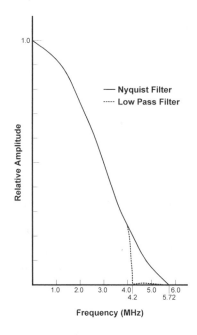

Figure 35.6 – Data Spectrum
After 100% Roll Off Nyquist Filter
+ Low Pass Filter

for error correction coding and overhead to facilitate robust reception, a net data rate of about 3½ MBytes per hour can be delivered through a standard broadcast television channel. In Pittsburgh, this can be doubled, if necessary to accommodate the utilization rate, by dedicating another line in the vertical interval to OnTV.

BROADCAST INTERNET

Beyond the delivery mechanism for bytes of data, the OnTV system deviates from NABTS. Instead a form of broadcast Internet data is used to create an Internet presence service. This is capable of transmitting the full range of multimedia and graphics that are available on the Internet. This can be text only, to save money and time, it can include color graphics pictures with image compression and full motion video with stereo sound, or it might include games that users can play.

An important feature of OnTV is local value added in which businesses can take pages in the service rather like the home pages available on the World Wide Web. These pages can list the services available from the businesses and provide other forms of advertizing. They can also include maps that show users how to reach the businesses owning the pages.

An adjunct to the OnTV service that can be accessed automatically when a user calls up a particular page is a supplemental (patented) mapping calculation routine that uses the known location of the receiver and location data transmitted with the page. In this instance, not only can the page show on a map where the business is located, but it can also highlight the route from the user's location to the business. In another variation of this application, the routine can make selections from a list so that a user seeking a particular service will only see those businesses that treat the user's location as part of their territories.

The data flow in the OnTV system recognizes that an intelligent processor (a personal computer) is at the receiving end of the channel. Messages are directed through the system to specific applications. Those applications have owners who are the businesses that support the service. The OnTV database is one such application. The OnTV database is stored in the user's PC or Mac; within the OnTV database, each service is unique. Storing of pages or other objects in the database happens on a recurring basis, with the refresh rates for those objects determined by the needs of the businesses that own them.

Storage in the OnTV database is intelligent and only stores in each user's copy items in which that user has expressed interest. If a user wanders into an unstored area, the user will be offered the chance to store what is there. If the user then requests something new, it will take the database some time to acquire it. How long will depend upon the frequency of recurrence, which, in turn, will depend upon payment choices made by the provider regarding how often to have it appear.

SYNERGY

One can easily foresee that, if services such as OnTV really take off with consumers, their need for bandwidth will grow rapidly. Until the day when their bandwidth hunger can be fed by new digital ATV transmission facilities, techniques such as those of WavePhore and Digideck may be able to supplant the NABTS vertical interval data channel with a bigger straw that can deliver more nourishment faster to a much larger data food chain.

36

Jitter Characteristics and Measurements

Transfering digital signals from one location to another or from one piece of equipment to another involves converting the signals into a physical representation in the analog domain at the sending end and then interpreting that representation to extract the data at the receiving end. This is necessary because any signals that are physically represented inherently have analog properties. Among these properties are the levels and timing of the data intervals and the transitions between them, the spectral distribution that results, and any signal distortions that occur in the various elements of the system through which the signal is sent. This is true whether the signals are modulated onto an RF carrier or transmitted directly as data using an appropriate form of encoding.

Among the analog effects that can alter digital signals are attenuation, spectral rolloffs and anomalies, overshoots, undershoots, time dispersion, and jitter. The first several are frequency- and amplitude-related effects while jitter is the sole timing-related disturbance. This chapter, we will examine the types of jitter in directly transmitted data signals, the methods for measuring each one, and some of the impacts they can have on system operation. We will also look into some of the system design approaches that can be used to minimize or mitigate the impact of jitter.

First published October, 1995. Number 33 in the series.

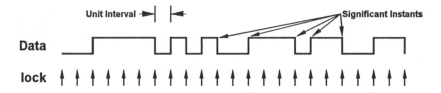

a) Relationship of Data & Clock for SDI Signals
Showing Unit Interval & Significant Instants

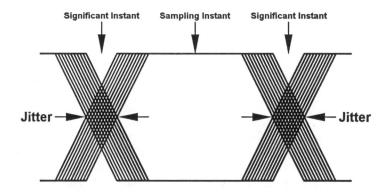

b) Eye Pattern Showing Jitter at Significant Instants

Figure 36.1 – Elements of Serial Data Signals & Relationship to Jitter

JITTER DEFINED

For this discussion, we will use the 270 Mbits/second (Mb/s) serial digital interface (SDI) signals (of SMPTE 259M) as an exemplar, but it should be recognized that the same concepts apply to everything from 1.3 Mb/s AES-3 audio data streams and 1.5 Mb/s MPEG-1 data streams through 1.5 Gb/s HDTV versions of the SDI. Only the absolute values used for equipment specifications and for measurements change when covering this large range of possible digital signals.

   In sending digital data streams from one place to another, data is encoded using one of several self-clocking methods that have been developed to optimize different performance characteristics. These include such schemes as NRZ, NRZI, AMI, block coding, bi-phase mark, and others. In general, the purpose of such methods is to minimize the DC and low frequency content of the signals and/or to minimize the bandwidth necessary to carry them. The important thing about all of these designs is that they allow the clock to be extracted from the data stream so it can be used to recover the data.

If the basic clock frequency of the digital signal is thought of as the carrier frequency of a radio signal, then jitter represents frequency or phase modulation of that carrier frequency. More precisely, jitter is the variation of a digital signal's **significant instants** (transition points) from their ideal positions in time. Jitter will cause the recovered clock and the data to become momentarily misaligned, with the result that the data will be misinterpreted when the misalignment becomes great enough. This, of course, will result in errors in the received data.

Jitter is defined and measured in terms of the **Unit Interval** (UI), which represents the period of one clock cycle and corresponds to the nominal minimum time between transitions of the serial data. This can be seen in Figure 36.1a, where the data of an NRZI signal and the related clock ticks are shown. Figure 36.1b shows the effect of jitter on the mid-point crossings of the data transitions, as would be seen on an eye-pattern presentation (repetitive display of transition points overlayed upon one another). It should be readily apparent that increasing jitter closes the eye and makes decisions between data states correspondingly more difficult, just as noise does in the amplitude dimension.

Returning to the frequency or phase modulation analogy, the amplitude of the jitter corresponds to the deviation of the carrier from its precise center frequency or of its phase from its unmodulated phase. The jitter similarly has a spectrum that corresponds to the frequency with which the modulated carrier varies around the center frequency or the unmodulated phase. Thus it is possible to plot an amplitude versus frequency characteristic of the jitter.

TYPES OF JITTER

Jitter of several types can be described and measured. In general, the differences have to do with the frequency of the recovered jitter modulation of the data signal's clock. Different frequency bands are used to describe the types of jitter. Also related to the descriptions of the various types of jitter are the methods for measuring them.

The very lowest frequency variations in the positions of a signal's transitions are termed **wander** and are not usually included in the measurements of jitter. There are, however, certain rather simple measurement techniques that measure **absolute jitter**, and these include all frequencies including those defined as wander. Wander is generally defined as being below a particular frequency, which is normally below 10 Hertz.

Jitter occuring at a rate greater than the highest frequency defined as wander is termed **timing jitter**. When jitter is measured relative to the transitions of a clock that has been extracted from the signal itself, it is called **alignment jitter**. As shown in Figure 36.2, timing jitter extends from the top of the wander spectrum to the top of the jitter spectrum being measured. Alignment jitter begins at a higher frequency that is dependent on the clock

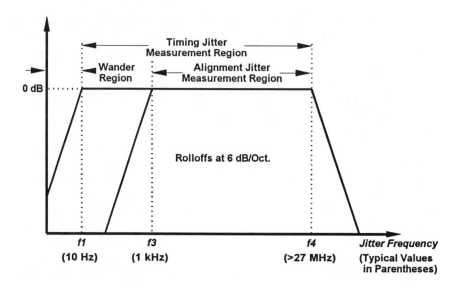

Figure 36.2 – Jitter Measurement Regions

recovery method and similarly extends to the top of the jitter spectrum being measured.

JITTER MEASUREMENTS

Several different instruments and connection arrangements can be used to measure jitter of various types. The more complex techniques will produce more informative results, but the simpler methods can be meaningful as a starting point for understanding the jitter occuring in a particular system.

The simplest method for observing and measuring jitter is the use of an oscilloscope with an external reference for triggering, as shown in Figure 36.3.

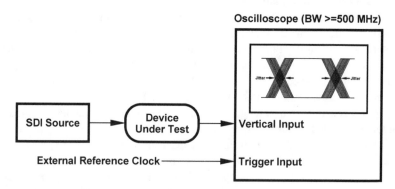

Figure 36.3 – Jitter Measurement with Oscilloscope
& External Reference Trigger

The bandwidth of the scope should be at least on the order of twice the clock frequency (≥ 500 MHz for a 270 Mb/s SDI signal). A digital storage oscilloscope is preferred to permit long measurements to be made using the infinite persistence capabilities of digital storage. The spread of the crossing points of an eye display will reveal the jitter amplitude. If a highly stable external reference, such as a clock derived from color black, is used, absolute jitter, including wander, will be displayed.

Greater detail regarding the characteristics of jitter in any particular case can be obtained by triggering the scope with a **clock extractor**. A clock extractor is a device that can recover the clock from an incoming data stream. Typically, it provides outputs of both the extracted clock and the serial data reclocked with the extracted clock.

An enhanced clock extractor, also called a **jitter receiver,** includes three sections and provides three outputs. The block diagram of a representative jitter receiver is shown in Figure 36.4. A clock extractor having only partial facilities can be quite useful. The first section has a passive loop through input (or power splitter), an equalizer, and a high bandwidth phase locked loop (PLL). The high bandwidth of the PLL results in the jitter of the input signal being reproduced in the extracted clock output (Output 1). The upper limit of the jitter bandwidth reproduced by the clock extractor (*f4* in Figure 36.2) is set by the loop bandwidth of this PLL. A frequency divider with a setable division ratio is provided prior to the output to permit oscilloscope measurement of jitter amplitudes greater than 1 UI (about which more shortly).

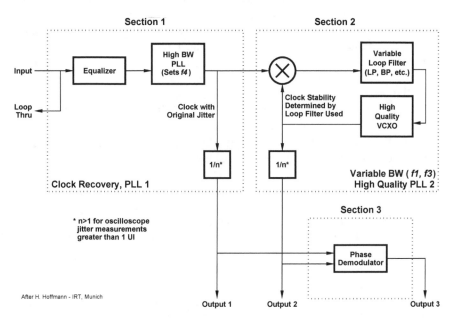

Figure 36.4 – Jitter Receiver Block Diagram

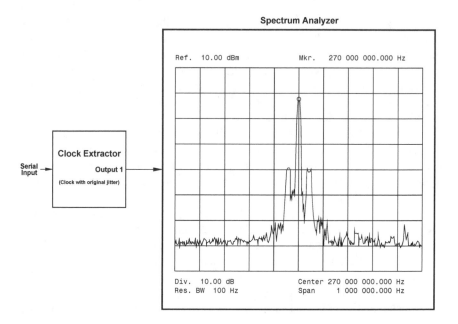

Figure 36.5 – Jitter Measurement Using a Clock Extractor
& Spectrum Analyzer

The second section of the clock extractor includes a second PLL that incorporates a variable bandwidth loop filter and a high quality (high Q) voltage-controlled crystal oscillator. The characteristics of the loop filter in this section determine the upper bound of the wander region (lower bound of jitter measurements — $f1$ in Figure 36.2) and the transition frequency between timing jitter and alignment jitter measurements ($f3$ in Figure 36.2). The output from this section (Output 2) also includes a frequency divider similar to that in the Output 1 path for oscilloscope measurement of jitter amplitudes larger than 1 UI.

The third section of a jitter receiver is a phase demodulator. This uses the outputs of the first two sections as inputs and demodulates the phase information of the clock with original jitter from section 1 with reference to the clock having stability set by the values of $f1$ and $f3$ in section 2. Output 3, from the third section, can feed a spectrum analyzer, a selective level voltmeter (effectively a tuneable receiver), or a filter and voltmeter arrangement to permit measurement of amplitude versus frequency characteristics. This arrangement is well suited for measuring the peak-to-peak jitter over the frequency bands specified for particular pieces of equipment.

Very useful measurements also result from use of a clock extractor in combination with a spectrum analyzer of one type or another. The exact configuration depends upon the capabilities of the spectrum analyzer used.

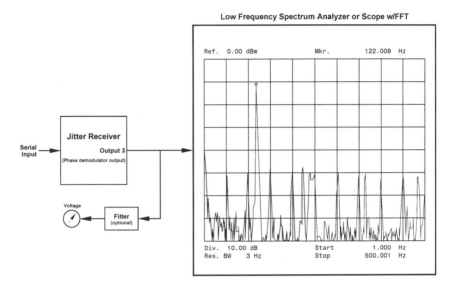

Figure 36.6 – Jitter Measurement Using Phase Demodulator Output
of Jitter Receiver

If a high frequency spectrum analyzer with a narrow resolution bandwidth is available, just the first section of the clock extractor can be used in combination with it to display the spectral characteristics of the jitter in terms of amplitude (deviation or phase change) and frequency. This is shown in Figure 36.5, in which the spectrum analyzer has been tuned to the clock frequency (270 MHz in this example) and in which two distinct sidebands are visible. These two sidebands represent a single modulating frequency (at about 40 kHz) that has added jitter to the clock somewhere in the system.

If a low frequency spectrum analyzer or an oscilloscope with a built-in FFT (Fast Fourier Transform) analyzer is available, the complete jitter receiver can be used in conjunction with it to create an amplitude versus frequency display similar to that just described. As shown in Figure 36.6, just the demodulated sideband energy is displayed. Depending upon the equipment used, this technique is good for measuring jitter up to about half the clock frequency. This may not be enough, as will be discussed below. Nevertheless, within its bandpass, it is suitable for discovering specific jitter modulation frequencies and their amplitudes.

The methods described so far are appropriate for oscilloscope jitter measurements in which the jitter is below 1 UI total deviation. When the jitter exceeds 1 UI and a scope is to be used, the dividers in jitter receiver Outputs 1 and 2 can be set to a division ratio higher than 1 and generally not higher than 10. This has the effect of reducing the frequency of the carrier by the division ratio used and of reducing the modulation deviation by a proportionate amount.

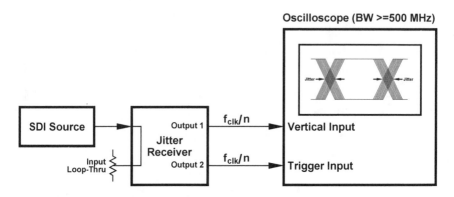

Figure 36.7 – Jitter Measurement of UI > 1 with Oscilloscope & Jitter Receiver – Set for Output Frequency Division

With proper selection of the division ratio, the result is that the phase rotation of the carrier does not exceed 360 degrees, and the eye does not close as it otherwise would at 1 UI. This setup is shown in Figure 36.7, wherein Output 2 is used for triggering a wideband oscilloscope and Output 1 drives the vertical channel. It is important to keep the division ratio as low as possible when making these measurements so as not to mask any word-related jitter effects, as might occur at a ratio of 10 and its submultiples.

EQUIPMENT CONSIDERATIONS

With all of these measurement methods available, for what are they used? First, they are used for characterizing equipment with respect to its jitter performance. Second, they are used to determine that system implementations do not permit accumulation of sufficient jitter to cause data errors or conversion distortions.

The jitter performance of equipment is characterized in several ways. **Input jitter tolerance** is the peak-to-peak amplitude of sinusoidal jitter that, when applied to an equipment input, causes a specified degradation of error performance. The input jitter tolerance of a piece of equipment can be specified through use of a template, as shown in Figure 36.8. The template specifies the minimum input jitter tolerance expected from the equipment, and the actual input jitter tolerance measured (also shown in Figure 36.8) will be <u>higher</u> than the template for properly operating units.

Jitter transfer is jitter that occurs on the output of equipment that results from jitter applied to the input of that equipment. The **jitter transfer function** is the ratio of the output jitter to the applied input jitter as a function of frequency. A template is used to specify the jitter transfer function in terms of **jitter gain** versus frequency. This is shown in Figure 36.9, where a compliant jitter transfer function is shown as falling <u>below</u> the template specification. The

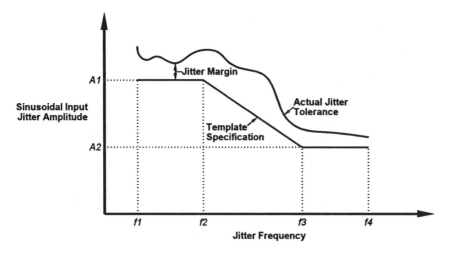

Figure 36.8 – Input Jitter Tolerance Template
& a Compliant Input Jitter Tolerance

template is specified and measurements are made from *f1* (the low frequency specification limit) to *fc* (the upper band edge of the jitter transfer bandpass).

Two types of jitter are measured identically at the output of a piece of equipment. **Intrinsic jitter** is the amount of jitter included in the output when

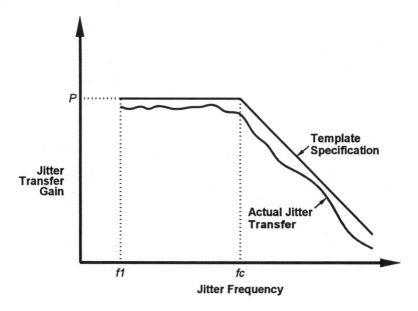

Figure 36.9 – Jitter Transfer Template
& a Compliant Jitter Transfer Function

the input to the equipment is jitter-free. **Output jitter** is the amount of jitter measured at the equipment output when the equipment is embedded in a network. Output jitter is a network specification, not an equipment specification. It comprises the sum of the jitter transfer of the jitter appearing at the equipment input and the intrinsic jitter of the equipment.

SYSTEM CONSIDERATIONS

In general, jitter accumulates as signals pass through a system. Good system design will assure that the output jitter of one piece of equipment does not exceed the input jitter tolerance of the succeeding equipment item. Indeed, some jitter margin should be allowed to account for variations in intrinsic jitter and jitter transfer performance with time and other variables. Periodically throughout a system and at any output analog interfaces, it may be necessary to incorporate a **jitter remover**; just as its name says, a jitter remover removes jitter.

When digital devices are cascaded, it is important to examine the characteristics of each to ascertain that the requirements of downstream devices will be met. It is relatively easy to envision a situation in which the output of a single device is satisfactory for a following device yet a cascade of similar devices is not. Consider the case of a device with a regenerator and with a critical overshoot in the performance of its PLL. This might result in the jitter transfer function shown in Figure 36.10 for a single device.

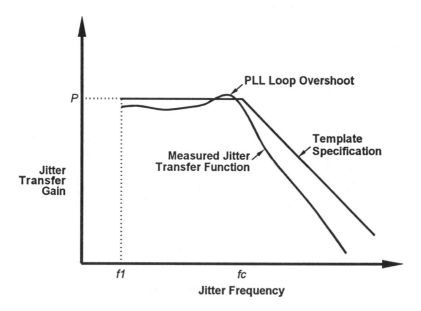

Figure 36.10 – Jitter Transfer Template
& a Non-Compliant Jitter Transfer Function

Cascading devices of this sort causes jitter growth which could exceed the input jitter tolerance of a following input. In this situation, avoiding cascades of equipment with similar overpeaking or overshoot would eliminate the problem, or jitter removers could be used more frequently in the equipment chain if such cascades were unavoidable.

In considering the performance characteristics of equipment to be incorporated into a system, it is important that investigations cover a sufficiently wide bandwidth. Thus the use of the phase demodulator measurements described earlier may be appropriate under certain operational conditions, but if problems arise, it may be necessary to apply different investigative techniques. This results from the fact that jitter frequencies as high as 20 per cent of the clock frequency (54 MHz in the case of a 270 Mb/s system, and occurring at twice the 10-bit word rate) have been known to disrupt proper operation of succeeding devices.

MITIGATING JITTER

It may be necessary to effectively eliminate jitter at various points in a system. This could be for reasons of an accumulation of jitter sufficient to exceed the input jitter tolerance of a succeeding device, as just discussed, or it could be to permit a high quality conversion to analog form. Jitter causes non-linearity in digital-to-analog (D/A) converters. High quality D/A conversion therefore requires removal of any jitter accompanying a signal prior to conversion, especially if the conversion clock is derived from the signal. For example, a jitter of <0.8 ns. (0.11 UI) will result in a 1 degree phase jitter of a 10-bit serial digital NTSC signal converted to analog. Similarly, for 16-bit audio performance, jitter under 0.1 ns is required.

A jitter remover works by converting serial digital signals back to parallel form, passing them through a first-in-first-out (FIFO) register, and then re-serializing them using a highly stable clock. As shown in Figure 36.11, the input to the jitter remover is structured very much like the input to any equipment equipped with an SDI input. This is followed by a relatively small

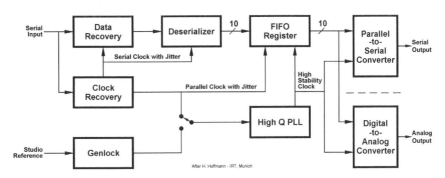

Figure 36.11 – Jitter Remover Block Diagram

FIFO with sufficient length to accommodate the longest time variations caused by the highest jitter amplitudes expected to be encountered. Finally, a high quality clock source drives either a serializer or a D/A converter to provide a virtually jitter-free output. Devices of this sort can be applied in a system as often as necessary to control jitter accumulation or to assure linear analog outputs. Care must be taken in implementing jitter removers to account for the additional delay they cause and to provide predictability of those delays.

CREDITS

The information that is included in this chapter is the result of over two year's effort by the SMPTE Working Group on Serial Digital Jitter, chaired by Ken Ainsworth of Tektronix. It had participation from a large number of individuals with an interest in making serial digital television an easily implementable reality. They completed two SMPTE Recommended Practices – on jitter specifications and jitter measurements, respectively. Their work will also form the basis for jitter specifications to be incorporated in all current and future SMPTE digital television standards.

In something of an experiment, this column will serve as the basis for an SMPTE Engineering Guideline to be published to help explain the functioning of the Recommended Practices. If the experiment works, we may try it again sometime.

Index

321